REPTILES

AND

AMPHIBIANS

IN COLOUR

by
HANS HVASS

Illustrations by
HENNING ANTHON

Translated and adapted by
GWYNNE VEVERS

BLANDFORD PRESS
LONDON

First English edition 1972
English text © 1972 Blandford Press Ltd,
167 High Holborn, London WC1V 6PH

ISBN 0 7137 0570 1

World Copyright Politikens Forlag A/S
8 Vestergade, Copenhagen, Denmark

A/598. 1'094

Colour sheets printed in Denmark by F. E. Bording A/S
Filmset in Photon Times 9 pt by
Richard Clay (The Chaucer Press) Ltd, Bungay, Suffolk
and printed in Great Britain by
Fletcher & Son, Ltd, Norwich

CONTENTS

PREFACE

Reptiles and Amphibians in Colour is designed as a handbook for the identification of the reptiles and amphibians, particularly those of northern, central and western Europe.

The colour plates illustrate all the species that are described, with the exception of a few lizards which very closely resemble those illustrated, both in size and general appearance. In many cases there is only one drawing for each species, but in certain cases two drawings are given, in order to illustrate differences between the sexes or between different colour phases. In other cases there are illustrations of the eggs and larval stages.

The second part of the book contains descriptive text. There is a short introduction to each of the five main groups: urodeles, anurans, chelonians, lizards and snakes, and an account of each species shown in the colour plates, as well as of a few others. Under each species the text is divided into four parts:

Identification: This gives the total length for the amphibians (urodeles and anurans), lizards and snakes, and the carapace length for the chelonians (tortoises and turtles). When talking of length measurements for amphibians and reptiles it should be stressed that, unlike birds and mammals, these animals do not stop growing when they have reached sexual maturity but go on growing throughout life. The figures given therefore refer to the average lengths of the adults. Measurements are given first in metric units as these are the most accurate; the conversions into inches should only be regarded as a rough guide. The measurements are followed by a description of the animal's general appearance, including coloration and pattern.

Distribution: This section describes the geographical range of the species, with in most cases a map showing the distribution within Europe.

Habitat: This gives the type of environment in which the species normally occurs.

Habits: This section gives general biological information, including type of home or shelter, breeding data, voice, diet, enemies and so on.

The original text is by Hans Hvass and the colour plates are by Henning Anthon. The distribution maps were drawn by Arne Gaarn Bak.

INTRODUCTION TO
REPTILES AND AMPHIBIANS

At the present time it is estimated that there are over 2,900 living amphibian species and over 6,000 living reptile species distributed throughout the world. To set these figures in perspective: there are about 4,500 living species of mammal, about 8,600 living species of bird, about 23,000 living species of fish and upwards of a million different insects. In Europe the amphibians are represented by only 43 species and the reptiles by 107 species; these figures refer only to living species.

The amphibians can be subdivided into three orders: caecilians (0), urodeles (19) and anurans (24). The figures in parenthesis give the number of European species.

Amphibians are poikilothermal ('cold-blooded') vertebrates with naked skin, without scales. Most of them breathe with gills in the larval stages, and with lungs when adult. The majority lay eggs which they do not tend, except in very few cases. The heart has three chambers, two auricles and one ventricle. Adult amphibians live on animal food, but some of the larvae are vegetarian.

The class Amphibia contains fewer species than the class Reptiles, but the individual numbers of amphibians are far greater than those of reptiles, birds or mammals.

Within the amphibians, the urodeles and the anurans stand so far apart that some authorities would classify them in separate classes. To do this, however, would make it difficult to find a place for the caecilians, a group of worm-like amphibians, so it is more appropriate to leave the classification as it is at the present time.

The reptiles are subdivided into five orders: crocodilians (0), chelonians (11), tuatara (0), lizard (63) and snakes (33). The figures in parenthesis again give the number of European species. Reptiles are poikilothermal vertebrates clad in horny scales or horny plates. They breathe with lungs and the majority lay eggs which are not tended. The heart usually has three chambers, two auricles and one ventricle which is, however, partly divided by an incomplete septum or dividing wall. The crocodilians have two completely separate ventricles, as in the birds and mammals.

The reptiles have evolved from long-extinct amphibians. Whereas the modern amphibians nearly all lay their eggs in the water, where there is no danger of desiccation, the egg-laying reptiles lay on land which means that the eggs must be protected against desiccation by a shell or similar structure. The newly hatched young reptile must be able to fend for itself

9

immediately and must therefore not be too small or under-developed. This means that the eggs must be relatively large in order to house sufficient yolk. The reptile's independence of water is associated with a skin which is horny and thick enough to prevent evaporation. This, in its turn, means that respiration through the skin can no longer take place, but reptiles show a greater development of the lungs and a more efficient breathing mechanism.

Most reptiles live on live food, seizing their prey in the jaws and swallowing it whole. With the exception of the chelonians (tortoises, terrapins and turtles) they have a number of peg-like teeth, which are adapted for seizing and holding prey but are not suitable for chewing.

In most reptiles the tail plays a part in movement. When limbs are present, they are usually small with the elbows and knees directed outwards. The toes have claws which may be used in climbing and digging, but not for seizing and holding the prey.

THE COLOUR ILLUSTRATIONS

1. **Smooth newt**

Triturus vulgaris (7–9 cm), male
a. male in breeding dress
b. female in breeding dress
c. larva

2. **Crested newt**

Triturus cristatus (12–16 cm), male in breeding dress
a. female
b. larva

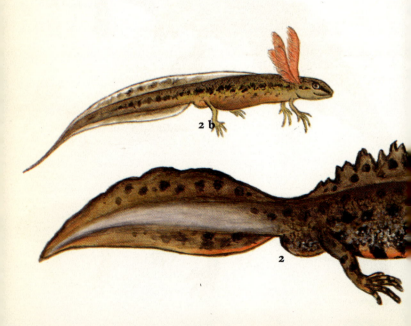

3. Marbled newt

Triturus marmoratus (13–14 cm), male in breeding dress
a. female

3 a

2a

4

4 a

4 b

5

5 b

5 a

4. **Alpine newt**
Triturus alpestris (7–12 cm), female
a. male in breeding dress
b. male, from below

5. **Palmate newt**
Triturus helveticus (8–9 cm), male in breeding dress
a. female
b. male in breeding dress, from below

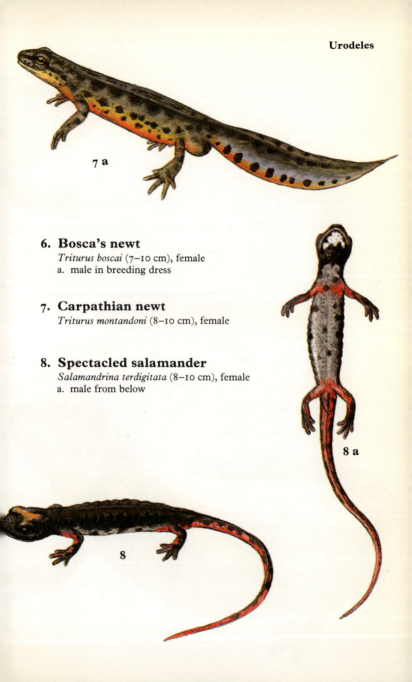

6. Bosca's newt
Triturus boscai (7–10 cm), female
a. male in breeding dress

7. Carpathian newt
Triturus montandoni (8–10 cm), female

8. Spectacled salamander
Salamandrina terdigitata (8–10 cm), female
a. male from below

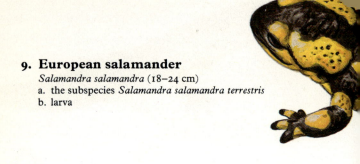

9. European salamander

Salamandra salamandra (18–24 cm)
a. the subspecies *Salamandra salamandra terrestris*
b. larva

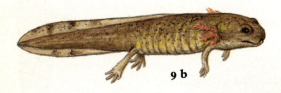

9 b

9

9 a

10

10. Alpine salamander
Salamandra atra (11–16 cm)

11

11. Gold-striped salamander
Chioglossa lusitanica (15–16 cm)

12

12 a

12. Sardinian mountain salamander
Euproctus platycephalus (8–11 cm), male
a. male, from below

13. Corsican mountain salamander
Euproctus montanus (8–11 cm), male

14. Pyrenean mountain salamander
Euproctus asper (10–16 cm), male
a. male, from below

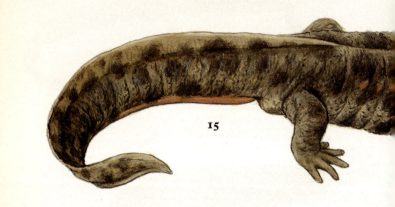

15

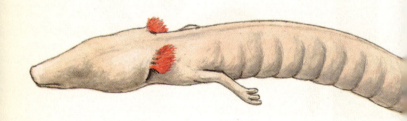

15. Pleurodele newt
Pleurodeles waltl (15–22 cm), male

16. Cave salamander
Hydromantes genei (10–12 cm)

17. Olm
Proteus anguinus (25–30 cm)

16

17

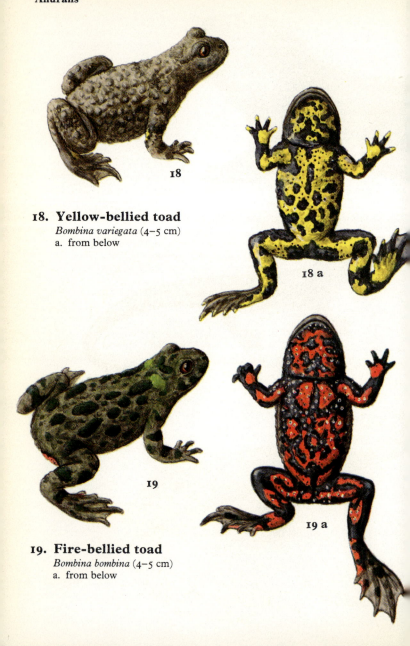

18

18. Yellow-bellied toad
Bombina variegata (4–5 cm)
a. from below

18 a

19

19 a

19. Fire-bellied toad
Bombina bombina (4–5 cm)
a. from below

20. Parsley frog
Pelodytes punctatus (*c.* 4 cm)

21. Painted frog
Discoglossus pictus (6–7 cm)
a. colour variant

20

21

21 a

22. Midwife toad
Alytes obstetricans (*c.* 5 cm)

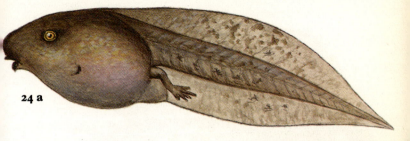

23. Iberian midwife toad
Alytes cisternasii (*c.* 4 cm)

24. Common spadefoot
Pelobates fuscus (6–8 cm)
a. tadpole
b. egg

25. Southern spadefoot
Pelobates cultripes (7–9 cm)

24 b

25

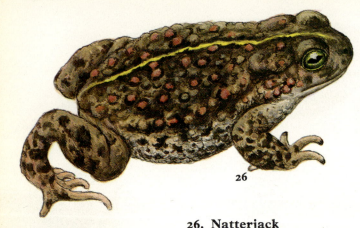

26

26 a

26. **Natterjack**
Bufo calamita (6–8 cm)
a. male showing vocal sac

27. **Green toad**
Bufo viridis (6–9 cm)
a. tadpole

27

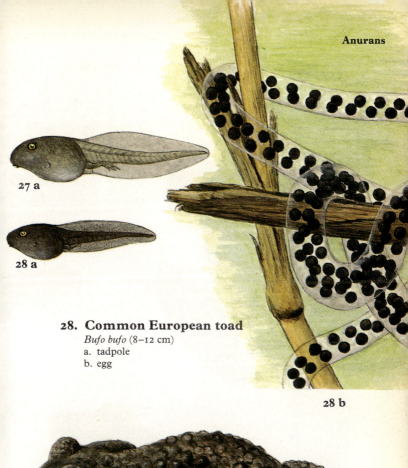

28. Common European toad

Bufo bufo (8–12 cm)
a. tadpole
b. egg

27 a

28 a

28 b

28

29 a

29

30 b

30 a

30 c

29. Field frog
Rana arvalis (6–8 cm)
a. with dorsal stripe

30. Common frog
Rana temporaria (7–9 cm)
a. egg
b. tadpole
c. tadpole metamorphosing

31. Italian agile frog
Rana latastei (4–7 cm)

31

32 b

32. Agile frog
Rana dalmatina (5–7 cm)
a. colour variant
b. tadpole

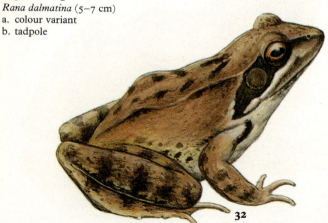

32

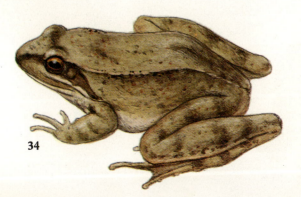

33. Spanish frog
Rana iberica (*c.* 6 cm)

34. Greek frog
Rana graeca (6–7 cm)

35

35 a

35 b

35. Edible frog
Rana esculenta (7–10 cm)
a. tadpole
b. newly metamorphosed frog

37 a

37

37 b

36. Marsh frog
Rana ridibunda (9–15 cm)

37. European tree-frog
Hyla arborea (3–5 cm)
a. male showing vocal sac
b. tadpole

36

38. European pond-tortoise
Emys orbicularis (12–25 cm)

39. Spanish terrapin
Clemmys caspica leprosa (18–20 cm)

40

40 a   41

40. **Hermann's tortoise**
Testudo hermanni (20–25 cm)
a. from behind

41. **Greek tortoise**
Testudo graeca (20–30 cm), from behind

42

42. **Loggerhead turtle**
Caretta caretta (60–90 cm)

43. **Hawksbill turtle**
Eretmochelys imbricata (45–60 cm)

43

44

44. Green turtle
Chelonia mydas (90–115 cm)

45. Leathery turtle
Dermochelys coriacea (150–270 cm)

46. Turkish gecko
Hemidactylus turcicus (9–10 cm)

46

47. Moorish gecko
Tarentola mauritanica (12–18 cm)

47

48. Naked-fingered gecko
Cyrtodactylus kotschyi (9–10 cm)

49. European gecko
Phyllodactylus europaeus (6–7 cm)

50

50. Mediterranean chameleon

Chamaeleo chamaeleon (25–30 cm)
a. with tongue extended

51. Slowworm

Anguis fragilis (30–50 cm)
a. variety with blue dorsal scales
b. new-born young

50 a

51 b

51

51 a

52

52. Grey burrowing lizard
Blanus cinereus (20–30 cm)

53. Fitzinger's lizard
Algyroides fitzingeri (10–12 cm)

54. Spinefoot lizard
Acanthodactylus erythrurus (18–20 cm)
a. juvenile

53

54 a

54

55. Algerian sand racer
Psammodromus algirus (20–27 cm), male

56. Spanish sand racer
Psammodromus hispanicus (10–12 cm)

57

57. **Round-bodied skink**
Chalcides bedriagai (11–12 cm)

58. **Sand skink**
Chalcides chalcides (24–40 cm)

58

59 b

59 c

59

59 a

59. Sand lizard

Lacerta agilis (15–20 cm), male
a. female
b. brown-backed variety
c. eggs

60. Viviparous lizard

Lacerta vivipara (10–18 cm), male
a. pregnant female
b. juvenile

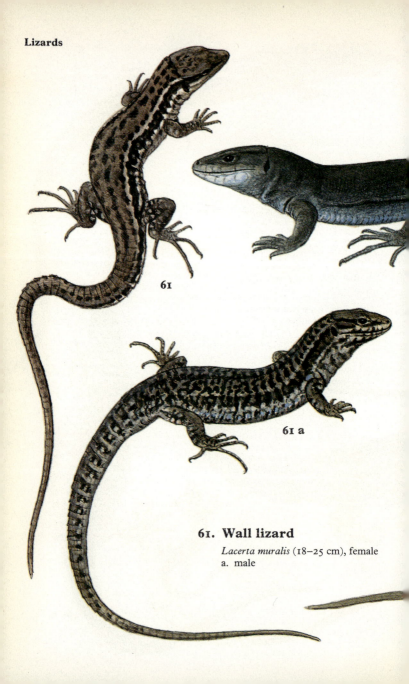

61

61 a

61. Wall lizard

Lacerta muralis (18–25 cm), female
a. male

62

62 a

62. **Ruin lizard**

Lacerta sicula (20–30 cm)
a. blue variety
b. self-coloured variety

62 b

63 b

63. Green lizard

Lacerta viridis (30–40 cm), male
a. female
b. juvenile

63

63 a

64. Schreiber's lizard
Lacerta schreiberi (25–30 cm)

64

65. **Eyed lizard**

Lacerta lepida (50–60 cm), male

65

66. European whip snake

Coluber viridiflavus (150–190 cm)

67. Horseshoe snake

Coluber hippocrepis (100–175 cm)

68

69 a

69

68. Leopard snake

Elaphe situla (80–110 cm)

69. Ladder snake

Elaphe scalaris (130–150 cm)
a. juvenile

70. Four-lined snake

Elaphe quatuorlineata (180–225 cm)
a. juvenile

70 a

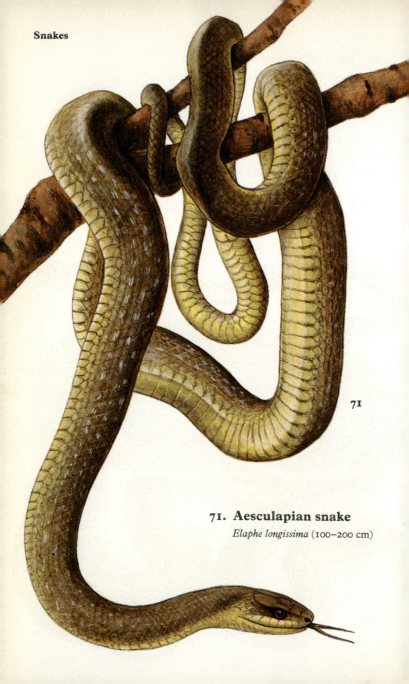

71

71. Aesculapian snake
Elaphe longissima (100–200 cm)

72. Smooth snake
Coronella austriaca (55–85 cm), male

72

72 a

73

73. Southern smooth snake
Coronella girondica (60–70 cm)

74. **Viperine snake**
   Natrix maura (80–100 cm)

75. **Dice snake**
   Natrix tessellata (100–150 cm)

76

76 c

76 a

76. Grass snake
Natrix natrix (70–150 cm)
a. black variety
b. cross-banded variety
c. eggs

76 b

77. **Hooded snake**
Macroprotodon cucullatus (40–50 cm)

78. **Montpellier snake**
Malpolon monspessulanus (100–200 cm)

79. **Snub-nosed viper**
Vipera latastei (50–60 cm)

79

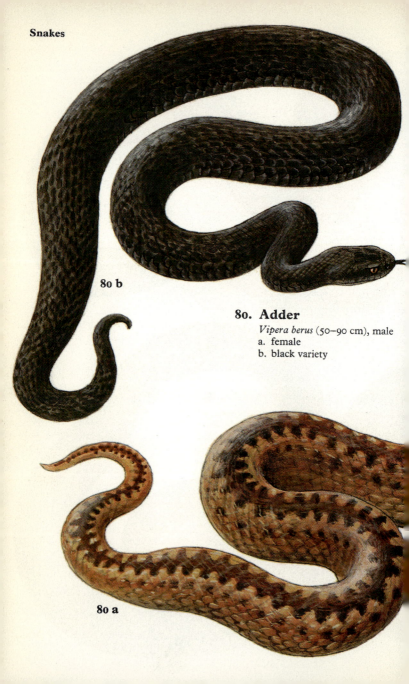

80 b

80. Adder
Vipera berus (50–90 cm), male
a. female
b. black variety

80 a

80

81. Meadow viper
Vipera ursinii (40–55 cm)

81

82

82. Asp viper

Vipera aspis (50–75 cm), female
a. male

82 a

83. **Sand viper**
Vipera ammodytes (60–90 cm)

83

DESCRIPTIONS OF THE SPECIES

Urodeles

The urodeles or newts and salamanders comprise some 225 species, of which only 19 are found in Europe. They are all rather similar to one another, and are slender, elongated amphibians with a short head and a long tail which often has a fold of skin along its upper and lower edge, which helps in swimming. The four limbs are of about equal length, the fore-legs usually with 4 toes, the hind-legs normally with 5. The thin naked skin plays an important part in respiration together with the ordinary breathing by the lungs. Most species have small teeth directed backwards, but these are so small as to be almost invisible. The young breathe by gills and some species retain these throughout life. Certain species have no lungs, respiration being effected through the skin and mouth. In reproduction there is no true mating but in certain cases the female is gripped firmly by the male and the cloacal openings are pressed against each other so that the sperm can be transferred to the female. More usually, however, the male produces one or more spermatophores which the female crawls over and takes up into her cloaca. There the sperm are released and they move up into the oviducts. The eggs which are surrounded by a coating of mucus are laid in the water, usually singly or a few together. Some urodeles produce live young. The larvae have three pairs of well-developed gills and at first have a tail but no limbs. Later, the fore-legs appear and finally the hind-legs. The changes which take place at the end of larval life—known as metamorphosis—are concerned mainly with the disappearance of the gills, when the lungs start to function.

1 Smooth newt
Triturus vulgaris

Identification: 7–9 cm ($2\frac{3}{4}$–$3\frac{1}{2}$ in.), exceptionally 11 cm (4 in.), the male larger than the female. Within its large area of distribution this species varies considerably in size, coloration and pattern. It is relatively slender and the laterally compressed tail, ending in a fine point, accounts for a good half of the total length. Outside the breeding season the skin is slightly warty and rather velvety but shortly after entering the water in spring it becomes smooth. During the breeding season the male has a crest of skin which starts at the back of the head and develops into a tall, undulating crest which attains its greatest height behind the root of the tail and then continues right along the top of the tail; there is also a similar crest along the lower edge of the tail. At this time the back toes of the male have narrow fringes of skin. The body and tail are marked with black spots which may be arranged in longitudinal rows. In the male the 'teeth' of the upper crest are nearly always black. The centre of the belly is orange and this colour continues past the hemispherical cloaca to the underside of the tail. During the breeding period the tail of the male has a lower stripe of mother-of-pearl iridescence. There are usually 5 dark longitudinal stripes on the head. After the breeding season the male goes on land and the crest and toe fringes are resorbed. The body becomes thinner and more angular, the tail rounder and the colours less intense, so that the animal is a uniform olive-brown on the

back, whitish-yellow on the sides and orange on the belly.

The female is more stoutly built with a plump belly. She has only a weakly developed ridge along the back in place of a crest, and there are no fringes on the back toes. The female is usually much paler, being greyish-yellow or brown. On the back there are nearly always two dark lines running parallel to the centre line and continuing out on to the tail. The orange colour on the belly is not so bright and in place of the male's relatively few dark markings the female has numerous dots. On the head there is often only a single dark line running through the eye.

Distribution: This is the most widely distributed urodele in Europe and by far the commonest in the greater part of its range. It extends farther north than any of the other species, reaching north of Trondheim in Norway and then eastwards through Sweden, Finland and Russia and far into Asia. Elsewhere in Europe it is found almost everywhere except in the Iberian Peninsula, the islands of the Mediterranean, Switzerland and the southern part of France.

Habitat: Lives in a wide variety of places, in woodland, heathland, marshes, mountainous areas and on plains, and in spring in all kinds of standing water. After the breeding season it moves on to the land, where it creeps about with the head and body raised and the tail flapping along the ground. In winter it hibernates under large stones, moss and tree roots, where it is protected from frost.

Habits: During the breeding season, which usually starts in early spring, the Smooth newt can be seen in large numbers in lakes, ponds, ditches and other places with standing water. In sunny weather it often floats spread-eagled at the water surface. When alarmed it swims down among the water plants at full speed, using the lateral undulations of the body and tail. At intervals it comes up to the surface to inhale air through the nostrils. The spent air comes out of the mouth in the form of small air bubbles which rise to the surface. The males are the first to be seen, and they are more numerous than the females. The male is very active during the breeding season, swimming around the female, undulating his dorsal crest, nudging, rubbing and nuzzling her with the tip of his snout, and this may go on for hours. Finally he produces a spermatophore (packet of sperms) which the female then creeps over and takes up. She lays large pale yellow eggs (about 3 mm or $\frac{1}{8}$ in. across) placing them singly on plants, often bending a small leaf round each egg as a form of protection. A female produces 200–300 eggs, but not all are fertilized and laid; some remain undeveloped in the ovaries. The eggs hatch in 2–3 weeks, and the pale yellow, almost transparent larvae emerge at the end of May. They have long-stalked, fringed gills and 2 short thread-like attachment organs under the head, with which they attach themselves to plants and rocks for the first few days. The limbs develop later, the hind ones being shorter and more powerful and these appear last. The larvae are very active and swim with jerky movements. They reach a length of 3–6 cm (1–2$\frac{1}{4}$ in.) and feed on fly larvae, water-fleas (*Daphnia*) and other small freshwater crustaceans. Larval life normally lasts 3–4 months and metamorphosis is completed in August. Sometimes the young remain in the larval stage until October, and exceptionally they may overwinter as larvae and continue their metamorphosis in the following spring. Immediately after metamorphosis, when the gills have been lost and the lungs have become functional the young leave the water and do not return to it until they are sexually mature—at an age of 2–4 years. Only exceptionally do they remain in the water after metamorphosis and spend the winter there, sometimes under the ice. There are records of fully grown adult specimens which have retained their gills and these are nearly always females.

Newts frequently shed their skin, a process known as sloughing; in spring this may occur once a week. The horny outer layer becomes loose so that the newt can creep out of it or pull it off with mouth or legs, leaving it hanging among the water plants. Sometimes the newt eats its own sloughed skin. Normally newts do not utter sounds, but if gripped around the body they may produce a weak sound, even though they lack vocal chords. They feed on *Daphnia* and other small crustaceans, snails, worms, larvae, tadpoles and insects, including those that fall in the water. When hungry they may attack members of their own species and bite pieces out of the tail. Newts have many enemies, including water-beetles, leeches, fishes, Grass snakes and other snakes, storks and other fish-eating birds.

2 Crested newt
Triturus cristatus

Identification: 12–16 cm (4$\frac{1}{2}$–6$\frac{1}{4}$ in.), exceptionally over 20 cm (7$\frac{1}{4}$ in.), the female being larger than the male. This is a relatively powerful species with a tail that is usually a little shorter than the body. The skin has numerous mucus glands and is very warty, particularly along the sides. The hind-legs are the strongest, and the male has slightly longer legs than the female. All the fingers and toes are free. The back and sides are dark brown or black, the sides with white dots, the underside orange or sulphur-yellow with a few large black spots. In spring the male has a tall toothed crest along the back, which starts at the back of the head and extends almost to the tip of the tail, with a break at the root of the tail. There is also a skin fold on the underside of the tail. The tail is laterally compressed and has an iridescent, mother-of-pearl band along each side, which is particularly conspicuous towards the tip. The female has no dorsal crest, but a slight dorsal ridge. She does, however, have a crest on the tail, but this is not so tall as in the male. The upperside of the female is usually darker, with smaller, less conspicuous spots. When egg-laying has finished the male loses the dorsal crest and the mother-of-pearl bands on the tail. The skin so to speak shrinks, leaving the newt with a more warty surface which is best seen in individuals that have moved on to the land.

Distribution: Almost the same extensive range as the Smooth newt but in the north extending only to about 61° N. Absent from Ireland, central and southern Italy, and Greece.

Habitat: Lives mostly in low-lying country and is only rarely seen at altitudes above 1,000 m (3,270 ft). Often found in woodland and during the breeding season it prefers very deep lakes with dense vegetation. Spends the day in deeper water than most other newts.

Habits: During the breeding season in spring the female is pursued by the male who places himself across her path and approaches so that their snouts touch each other. The male usually holds his legs stiffly with the body free of the ground and arches his back. He bends his tail towards the female and slaps her flanks with it. He then stretches out flat on the bottom and with his body twitching releases a spermatophore. The female crawls over the spermatophore, which becomes attached to her cloaca and the sperms pass up into her body. This procedure may be repeated several times until the male swims away and later starts to court another female. Egg-laying usually begins in April–May and continues for a long time.

The 200–300 eggs are laid singly, each being carefully attached to the underside of a small leaf which is curled round it. After about 2–3 weeks the eggs hatch into greenish-yellow larvae (length 8–10 mm or $\frac{3}{8}$ in.) which hang motionless from water plants for the first few days. The larvae feed voraciously on insects and worms. Metamorphosis is complete in September when the larvae are 6–8 cm ($4\frac{1}{4}$–3 in.) long and have lost their gills and acquired functional lungs. The majority then go on land to search for winter quarters under moss or tree roots or in holes in the ground. They become sexually mature after 2 years and again move down to the water. The larvae which hatch late and do not complete their metamorphosis spend the winter in the water. Some of the adult newts remain in the water until the autumn and a few spend the winter in the mud at the bottom of lakes, but the majority hibernate on land. Some of the skin glands produce a secretion which has an irritating and possibly lethal action on predators. When disturbed these newts may produce a very pungent smell. They feed on worms, snails and larvae, and the prey is usually shaken thoroughly before being swallowed.

3 Marbled newt
Triturus marmoratus

Identification: 12–16 cm ($4\frac{1}{2}$–$6\frac{1}{4}$ in.). Very similar to the Crested newt. Snout short and rounded, eyes very large, and the tail which is half the total length is strongly compressed laterally. Skin granular with numerous pores on the head and at the base of the limbs. The male's untoothed dorsal crest is very tall and is separated from the tail crest by a short gap. Outside the breeding seasons these crests become considerably reduced but do not disappear completely. The body of the female has no dorsal crest but has a shallow groove along the middle line of the back.

The body is grass-green attractively marked with irregular dark spots and marbling. The ground colour varies from yellow-green to dark olive-green, and is more intense when the newts are on land. In the breeding male the dorsal crest has black and whitish vertical bands and there is a conspicuous silvery stripe on each side of the tail. The females and young have a narrow orange stripe running along the back.

Distribution: Found throughout the whole of the Iberian Peninsula and in a large area of France west of a line running from the Seine through Paris and Lyons to the Rhône; also occurs

just east of the Rhône along the Mediterranean coast.

Habitat: Mainly in low-lying country and usually not at altitudes above 400 m (1,300 ft). Not very demanding as regards habitat, and thrives equally well on sand dunes, clay soil and limestone hills. For breeding it seems to prefer quite small lakes, ponds or ditches with dense vegetation.

Habits: Spends the breeding season in the water, sometimes remaining until far into the summer. On land it is nocturnal, first emerging at twilight to search for food. Breeding starts early in the spring and the courtship is more or less the same as in the Crested newt. Egg-laying takes place in March–May, depending upon the temperature, and the female produces 200–300 eggs which she attaches singly to the leaves of water plants. The eggs hatch in 2–3 weeks and the larval stage lasts for about 3 months. In that part of France where both the Marbled and the Crested newt occur it is not uncommon to find hybrids. These usually resemble the Marbled newt on the back and the Crested newt on the underside and they were at one time regarded as a separate species. The diet of the Marbled newt is similar to that of the Crested newt.

4 Alpine newt
Triturus alpestris

Identification: 7–12 cm ($2\frac{3}{4}$–$4\frac{1}{2}$ in.), the male usually 7–8 cm ($2\frac{3}{4}$–3 in.), the female 9–11 cm ($3\frac{1}{2}$–$4\frac{1}{2}$ in.), exceptionally 12 cm ($4\frac{1}{2}$ in.). The skin has fine granules but becomes smoother when the newts are living in the water during the breeding season. The greyish-black colour of the back

changes to bluish during breeding, and the sides become pale yellow with black spots, this area being bordered below by a sky-blue stripe, which is particularly noticeable in the male. In both sexes the belly is permanently bright orange-red. During the breeding season the male also has a low yellowish dorsal crest, regularly spotted with black. The female is less brightly coloured than the male and acquires small white warts along the sides when she moves on to the land.

Distribution: The main range covers the mountain regions of central Europe, that is the mountains of France and Germany, the Alps, Carpathians, the northern part of the Apennines and the mountains of the Balkans. Also found from southern Jutland in the north to Greece in the south and from Spain in the west to Rumania and Bulgaria in the east, as well as in a few places in Russia.

Reaches higher altitudes than any of the other European newts, often being found at 1,000 m (3,270 ft), and exceptionally up to 3,000 m (9,800 ft), but it also occurs in low-lying country.

Habitat: In lakes and ponds, preferably with clear water, but some wander over great distances, and are sometimes found far from water, under rocks and tree roots.

Habits: Breeds earlier in the year than most other urodeles, in certain regions immediately after the water has become free of ice. Courtship usually occurs in March–April. The male starts by approaching the snout and cloaca of the female, coming alongside her and vigorously flapping his tail. The eggs are usually laid in April–May, and are attached to water plants, either singly or in small groups. The female lays about 150 eggs, and these hatch in 2–4 weeks, depending upon the temperature. The dark grey larvae are 7–8 mm ($\frac{5}{16}$ in.) long on hatching, and they grow fast and acquire large, bushy external gills. The larval stage usually lasts about 3 months, but it is not uncommon for larvae to over-winter without metamorphosis, and sometimes they may even breed in the larval stage, a phenomenon known as neoteny which occurs in certain other urodeles.

Albino specimens of the Alpine newt have also been found, mainly in alpine lakes at high altitudes. This species does not usually swim about in open water, but keeps mostly near to the bottom, whether it is close to the edge of a shallow pond or out in the middle of a deep lake. The diet consists of small crustaceans, insects, larvae and worms. The sense of smell is evidently better than in other urodeles, and there is indeed a greater need for this as Alpine newts take most of their food from the bottom or down in the mud and do not usually hunt free-swimming prey.

5 Palmate newt
Triturus helveticus

Identification: 8–9 cm (3–3$\frac{1}{2}$ in.), the female exceptionally up to 10 cm (4 in.). Body slender and the tail slightly longer than half the total length. Skin almost completely smooth to the touch, although it has very fine granulations. During the breeding season the male has a short filament (5 mm or $\frac{1}{4}$ in. long) at the end of the tail which in other respects ends rather abruptly. At this time he has a fairly low dorsal crest which is directly continuous with the tail crest. The male also has a slightly raised ridge along each side of the back and well-developed webs on the hind-feet.

The upperside of the male is olive-green or olive-brown, with dark green markings and sometimes a faint golden sheen. On each side of the head there is a dark stripe which starts on the snout and runs through the eye. The coloration of the female is similar to that of a female Smooth newt. In both sexes the underside is pale orange or yellow, usually uniformly coloured, but sometimes with brown spots. The body is darker when the newt is on land, and along the middle of the back there is a narrow yellow or reddish stripe, which is particularly noticeable in the female.

Distribution: England, southern Scotland, France, Holland, Belgium and western Germany eastwards to a line running from Hamburg southwards to northern Switzerland. Also found in northern Spain and in Portugal southwards to Oporto. Prefers mountainous regions but can also be found in low-lying country, especially where the soil is sandy, calcareous or peaty. In the northern part of the range this species is found at altitudes up to 1,000 m (3,270 ft), in the southern part up to 2,000 m (6,500 ft).

Habitat: Like most other urodeles, spends the spring in the water and the remainder of the year on land. Not very particular as to the type of water, for it can be found in small ponds, large lakes, slow-flowing streams and rivers and even in brackish water. Prefers shallow water, and in large lakes therefore lives mainly along the shores. The favourite habitat is a gently sloping muddy bottom with dense vegetation.

Habits: Emerges early from its winter quarters, usually in March–April, but in the more southerly localities in February, and moves straight into the water to breed. The male's approaches to the female are quite violent. Time and again he positions himself in front of her, snout to snout, bends his whole body towards her and lashes his tail. Over a period of 3–4 weeks the female lays 300–400 eggs, which are attached singly to aquatic plants. The newly hatched larvae are 8 mm ($\frac{5}{16}$ in.) long, and metamorphosis is completed when they are about 25 mm (1 in.); this is normally in July–August, but the larvae sometimes overwinter under the ice and only finish their metamorphosis in the following year. Palmate newts feed mainly on small worms, insects and their larvae.

6 Bosca's newt
Triturus boscai

Identification: 7–10 cm ($2\frac{3}{4}$–4 in.). Very similar to the Smooth newt but without a dorsal crest and the tail ends in a short filament. In the male the upperside is yellow-brown, in the female dark olive-brown, with small dark markings. There is often a yellow-brown dorsal stripe and the upperside is separated from the orange-yellow belly by a whitish or yellowish stripe along each side.

Distribution: Found only in the Iberian Peninsula and most commonly in the western part of Galicia in the north, through Portugal to the Algarve in the south; also in the western and central parts of Spain.

Habitat: Lives in cool, clear mountain lakes and streams during the spring, but by June it has moved on land where it spends the rest of the year.

Habits: Swims slower than the commoner newts, having a more weakly developed crest. Feeds exclusively on living food in the form of small crustaceans, insects, larvae and worms.

7 Carpathian newt
Triturus montandoni

Identification: 8–10 cm (3–4 in.). In the male the tapering tail ends in a thin filament which disappears when the newt leaves the water. The female's tail is relatively longer but has no filament at the end. In the male the upperside varies from pale yellow through brown to grey or olive-green, and is variously marked with spots and marbling which may form longitudinal stripes. During the breeding season the male has a low dorsal crest extending from the back of the head to the tip of the tail. The underside is orange with few or no markings. The female is usually paler, sometimes almost completely uniform in colour.

Distribution: In the Carpathians and Tatra Mountains in central Europe.

Habitat: Usually at the foot of the mountains but can be found at altitudes up to 800 m (2,600 ft).

Habits: At the end of March or beginning of April moves into small forest pools to breed, and returns to land in the middle of June. When the autumn rains start it is said that some adult members of this species go back to the water and spend the winter in the mud at the bottom, but it is not clear whether they all do this or whether some spend the winter on land. Feeds on worms, small crustaceans and larvae.

8 Spectacled salamander
Salamandrina terdigitata

Identification: 8–10 cm (3–4 in.). Slender with an almost cylindrical tail. Only 4 toes on each foot. The popular name refers to the reddish-yellow mark framing the very large eyes. Upperside dull black, underside pale with black spots. Underside of tail and inner parts of limbs fiery red.

Distribution: Found only in Italy from Liguria in the north to the Naples area in the south, the distribution being limited to the east by the western slopes of the Apennines. Most abundant in the region around Genoa.

Habitat and habits: Lives mostly under rocks and moss along streams and rivers, only going into running water in the breeding season. The female lays in early spring, placing the eggs in small groups on rocks on the bed of a stream.

Feeds on small insects which it catches with its long, protrusible, sticky tongue.

9 European salamander
Salamandra salamandra

Identification: 18–24 cm (7–9½ in.). Body rather powerfully built with a broad head and an almost completely cylindrical tail. On each side of the head there is a large swelling formed by glands, known as the parotoids; these have conspicuous pores. There is also a double row of gland pores running along the back. The glossy skin is black, marked on the upperside with large irregular yellow or orange spots

at almost any time of the year but mainly occurs in spring or early summer. During the mating period this species produces a characteristic scent which is probably important in enabling the sexes to find one another. The male chases the female, thrusts at her with his snout and sometimes bites her. He then pushes himself under her body and with his back up against her grips her firmly with his fore-limbs while he releases a spermatophore. Sometimes he may twist round so that the spermatophore can be taken up directly by the female. The young are not born for several months, and usually not until the following year. The female goes down to a pond or stream and crawls out a short distance into the water, but without becoming completely submerged. She gives birth to 12–20 or sometimes 50 young. These are 2–3 cm ($\frac{3}{4}$–1 in.) long and brownish with large external gill tufts and they can swim immediately with the help of the well-developed tail. These larvae feed on *Daphnia* and other small crustaceans and when they are 6–7 cm ($2\frac{1}{4}$–$2\frac{3}{4}$ in.) long and about 3 months old they metamorphose and go on land. The males never return to the water and the females only do so to give birth.

European salamanders feed on worms, insects and larvae and also consume many slugs. Evidently the slime produced by the slugs does not annoy the salamanders, although they can be seen to dry their mouths on grass or earth after such a meal.

and bands, which vary considerably from one individual to another.

Distribution: The species is subdivided into some 10 forms or subspecies, of which 8 occur in Europe, throughout the whole of southern and most of central Europe. The eastern limits of the range extend from Poland to the Black Sea. Absent from Britain.

Habitat: Lives entirely on land in shady, damp places—in holes in the ground or under tree stumps—mainly in hilly regions with deciduous forest, and sometimes up to altitudes of 1,000 m (3,270 ft).

Habits: Usually only active at night or in twilight, but during or after rain it may come out during the day. Like so many other amphibians it produces a poisonous secretion with irritant properties, which may even be lethal to small animals. Apart from the period of hibernation, mating may take place

10 Alpine salamander
Salmandra atra

Identification: 11–16 cm ($4\frac{1}{4}$–$6\frac{1}{4}$ in.). Skin glossy and uniformly coal-black.

Head flat with a short broad snout and the parotoid glands are very large. Along the back there is a row of large gland pores and there is also a row on each side of the body, which has the appearance of being subdivided by distinct furrows which continue out along the almost completely cylindrical tail.

Distribution: Found in the Alps at altitudes of 300–3,000 m (900–9,800 ft), the range extending into neighbouring areas of France, Germany, Italy, Switzerland, Austria, Hungary, Yugoslavia and Greece.

Habitat: In damp places, but never enters the water. Usually lives under stones or tree roots.

Habits: Usually only active at night, but after heavy rainstorms may be seen by day, often in very large numbers. Mating takes place in July–August, the male lying along the back of the female and holding her firmly by his fore-limbs while releasing his spermatophore. The period of gestation is about 1 year at an altitude of 1,000 m (3,270 ft) and probably longer (2–3 years) at 3,000 m (9,800 ft). The female produces 10–20 eggs in each ovary, but only 1 or 2 eggs in each ovary develop into young, which are 4–5 cm (1½–2 in.) long at birth. Their gills have disappeared before birth. The remaining eggs break up into a mass which provides food for the embryos that do develop.

Alpine salamanders feed mainly on earthworms and slugs.

11 Gold-striped salamander
Chioglossa lusitanica

Identification: 15–16 cm (6–6¼ in.). Very slender with glossy skin. Body and tail almost completely cylindrical except for the outermost part of the tail which is laterally compressed. The tail is almost twice as long as the head and body. Tongue very long and attached at the front; it can be protruded a considerable distance. Upperside brown or black with 2 longitudinal orange or coppery bands, which fuse to form a single band on the tail. Underside grey or pale brown.

Distribution: Restricted to north-western Spain from Galicia to Old Castile and northern Portugal, but only in isolated colonies.

Habitat: Prefers wooded hill country with streams and springs, at altitudes up to 400 m (1,280 ft).

Habits: Much faster and more active than most other urodeles and is sometimes seen in the middle of the day. Usually, however, it does not emerge from under moss and stones until the evening. As in many lizards, the tail can be broken off or autotomized and a new one will quickly grow out again.

Little is known about the breeding conditions. Mating may possibly take place on land but the eggs are laid in the water, usually in springs and streams. This species feeds mainly on flies and other small insects which are caught with the long, sticky tongue.

12 Sardinian mountain salamander
Euproctus platycephalus

Identification: 8–11 cm (3–4¼ in.). Body slender with a strikingly flat head and a long snout, which resembles that of a pike. Tail strongly compressed laterally. In the male the hind-feet have a spurlike outgrowth which looks rather like a sixth toe. The upperside of the body is brown, nearly always with a pale yellowish or brownish stripe along the back and with dark spots or marbled markings. Underside yellowish or whitish with black spots.

Distribution and habitat: Found only in Sardinia, high up in the mountains, particularly in the area of Monte Gennargentu where it occurs at altitudes up to 1,800 m (5,870 ft).

Habits: Said to breed twice a year, in late spring and early autumn. Breeding takes place in the water, usually in among rocks on the edges of lakes and streams. The male approaches the female, crawls in under her and curls his tail round her body in front of her hind limbs so that she cannot escape. He holds on firmly for at least 1 hour, and sometimes for over 24 hours. During mating a spermatophore is transferred to the female and the eggs are laid shortly afterwards. This salamander moves slowly both on land and in the water and feeds on slow-moving animals such as earthworms, larvae and slugs.

Also known as the Flat-headed salamander.

13 Corsican mountain salamander
Euproctus montanus

Identification: 8–11 cm (3–4¼ in.). Slender with a short snout. The rear part of the head appears broad on account of the large parotoid glands. Tail only laterally compressed in its outer third. Upperside brown or olive-grey with dark spots, and usually with a yellowish-brown dorsal stripe. Underside grey or brown, often with dark spots.

Distribution and habitat: Only found in Corsica where it lives in or near mountain streams at altitudes of 700–2,500 m (2,300–8,200 ft).

Habits: Breeding and diet as in the preceding species.

14 Pyrenean mountain salamander
Euproctus asper

Identification: 10–16 cm (4–6¼ in.). A large robust salamander with a rough skin. Upperside grey to greenish-black, usually with a lemon-yellow dorsal stripe which extends out on to the tail. Underside orange-red in the female, yellowish in the male.

Distribution and habitat: In the Pyrenees, particularly in the central and eastern parts. Like the related salamanders this species lives high up in the mountains at altitudes of 700–3,000 m (2,300–9,800 ft), sometimes so high that it is only active for 4 months in the year.

Habits: Usually breeds in June, in much the same way as the 2 preceding species.

15 Pleurodele newt
Pleurodeles waltl

Identification: 15–22 cm (6–8¾ in.), sometimes 20–30 cm (8–12 in.). This is the largest urodele in Europe, exceptionally reaching a length of 40 cm (15¼ in.). Along each side of the back there is a row of wart-like knobs produced by the ends of the ribs; these have sharp points which may protrude from the warts. The body is powerfully built, the head broad and toad-like, and the tail has a narrow keel. The ground colour is grey-brown or grey-green, with dark spots over the whole. The warts on the sides are usually orange.

Distribution: Southern and western areas of the Iberian Peninsula, and also in parts of Morocco.

Habitat: Lakes, ponds, bogs and ditches with dense vegetation, where it can shelter on the bottom. Seldom comes on land and if the water dries up in summer it digs down into the mud and aestivates.

Habits: A very voracious species, which eats anything it can overpower, from small insects to large worms, as well as tadpoles and small frogs, and even members of its own species. It can probably breed at almost any time of the year. The female lays 200–300 eggs at a time and fastens them singly or in small groups to stones and water plants. She lays 4 or 5 times a year.

16 Cave salamander
Hydromantes genei

Identification: 10–12 cm (4–4½ in.). A slender species with smooth glossy skin, a broad head, a relatively short, cylindrical tail, 11–13 vertical furrows along the sides of the body and short webs on the fingers and toes. It has no functional lungs and breathes solely through the skin and the vascularized lining of the throat. The ground colour is usually dark brown which is marked with large and small irregular reddish-yellow spots, but there is considerable variation.

Distribution: The Maritime Alps in south-eastern France, north and central Italy southwards to the Abruzzi Mountains and also in Sardinia.

Habitat: Lives most of the time in damp places, underground holes or in rock crevices and only emerges at night, particularly when it is raining.

Habits: Feeds on insects and spiders, which it catches with the long, protrusible tongue. Probably breeds mainly during the spring, the female giving birth to live young which are about 3·5 cm (1⅓ in.) long.

17 Olm
Proteus anguinus

Identification: 25–30 cm (10–12 in.). Body eel-like, head elongated with a flat snout and rudimentary, non-functional eyes which are overgrown with skin. In the neck region there are three pairs of bushy external gills which appear red due to the blood circulating within them. Tail laterally compressed with a narrow keel above and below. Limbs very small with 3 toes on the fore-feet and 2 on the hind-feet. Skin almost completely white, but it darkens when the animal is kept in the light.

Distribution and habitat: Lives in underground lakes and streams near Trieste, in Istria and in north-western Yugoslavia; also in the extreme south of Austria.

Habits: Entirely aquatic, never coming on land; the water in which it lives has a temperature of 5–10°C (41–50°F). The Olm does not undergo metamorphosis and retains the larval gills throughout life. The female normally produces only one or two live young at a time, but these are 10 cm (4 in.) long at birth. Feeds mainly on small crustaceans which it finds in the mud.

Anurans

There are about 2,600 living species of anurans (frogs and toads) in the world, of which only 24 are found in Europe. They are all fairly similar to one another. The body is short and the hind-legs, which are much longer and more powerful than the fore-legs, are used for jumping and also for swimming. In contrast, the larval stages, here known as tadpoles, move with the help of a fin which runs along the back as well as above and below the tail. The naked skin is smooth or warty and together with the lungs it plays an important part in respiration. The young live in the water but in most species the adults spend almost all their lives on land, but usually in fairly damp situations as the skin cannot withstand desiccation. In winter anurans hide away and hibernate. There is no true mating, since the male has no copulatory organ. Before and during egg-laying the male clasps the females with his fore-legs and sheds sperms on to the eggs after they have been laid.

When the tadpoles hatch from the eggs they attach themselves to water plants by a pair of suckers. On the sides of the head they have bushy external gills, and the tail is long and laterally compressed. Within a few days the tadpoles change shape, the head and body become thicker, and the external gills disappear and are replaced by internal gills, which have developed inside the gill clefts. Later on, the hind-legs grow out first and these are followed by the fore-limbs which break out through the gill-covers. The gill opening closes, and finally the tail disappears completely. The gill-breathing, vegetarian tadpole has changed or metamorphosed into a lung-breathing predatory frog or toad.

18 Yellow-bellied toad
Bombina variegata

Identification: 4–5 cm (1½–2 in.). One of the smallest European anurans, with a short toad-like body and a broad head with rounded snout. The back has a large number of prominent warts, but the parotoid glands are not conspicuous. In the male there are small black horny outgrowths on the dorsal warts which are, however, so small that they cannot be seen with the naked eye. The webs extend to the tips of the toes. Back grey-brown, olive-green or yellowish with a bronzy sheen. Underside varies from straw-coloured to orange with grey-blue or black spots and marbled markings. In some specimens the coloration of the belly is reversed, with yellow spots on a dark background. The yellow colour extends from the underside on to the fore- and hind-legs. The tips of the fingers and toes are completely yellow. The distribution of the colours varies considerably between individuals and is dependent upon age and locality. The male has no vocal sac. During the breeding season he has small, black, horny tubercles on the underside of the lower arm, on the insides of the first, second and third fingers and also on the undersides of the second, third and fourth toes.

Distribution: Central and southern Europe, but absent from Brittany, the Pyrenees, Spain and Portugal. The northern limit runs from the Netherlands south-eastwards along the Carpathians to the Black Sea. Not found on the islands of the Mediterranean or in the Peloponnese. In many parts of the range the distribution is very local.

Habitat: Occurs both in low-lying country and in mountainous regions and in the Alps it extends up to altitudes of 1,500 m (4,900 ft). Lives equally well in clear or muddy water, with or without vegetation and even in quite small pools. In winter it hibernates in holes in the ground or under rocks.

Habits: Usually emerges from hibernation at the end of April and often covers very long distances before finding a satisfactory pond. In spring and summer evenings or after a rain shower one can hear the male's deep, but quite soft calls. This species may breed two or three times in the year and eggs can be found from the beginning of May to the middle of July and sometimes right into September. The female lays about 80–100 eggs, either singly or in small groups of not more than 10 eggs. The newly hatched tadpoles are 6–7 mm ($\frac{1}{4}$ in.) long, but before metamorphosis they will have reached a length of 4–5 cm ($1\frac{1}{2}$–2 in.). This and the following species are very similar to one another in appearance and habits, and in places where both species occur it is not uncommon to find hybrids.

19 Fire-bellied toad
Bombina bombina

Identification: 4–5 cm ($1\frac{1}{2}$–2 in.). Body proportions similar to those of the preceding species. Male somewhat smaller than the female, with a broader head and larger fore-limbs. Back very warty, but belly almost completely smooth. The toes on the relatively short hind-legs have webs which are particularly well developed in the male. During the breeding season the male has an area of small, dark, horny tubercles on the inside of the lower arm and of the first and second fingers. The back is usually greyish olive-green with dark brown or bottle-green spots which together form a symmetrical pattern, and usually with two pale green markings between the shoulders. Underside bluish-black with large orange-red spots and numerous small black dots surrounded by white rings; the colour pattern continues out on to the surfaces of the hands and feet. The coloration varies from individual to individual and is also dependent upon locality, age and season. The male has well-developed vocal sacs, which are inflated when he is croaking.

Distribution: Mainly in eastern Europe. The western limits run along the Weser and south-eastwards along the Danube to the Black Sea. The northern limit runs from southernmost

Denmark eastwards along latitude 55° N to far within the Soviet Union. In most areas the distribution is quite local.

Habitat: Lives in low-lying country, and in contrast to the Yellow-bellied toad does not occur in the mountains. Prefers lakes and ponds with dense vegetation and clear water but may sometimes also be found in muddy pools. Late in the year it leaves the water and enters hibernation in a hole in the ground, in a stone wall or under the root of a tree.

Habits: Usually emerges late in the spring and goes straight down to the water. The male often lies at the surface with limbs outstretched, with the body and vocal sacs inflated, and with only the snout and eyes above water; the throat is much distended by the vocal sacs. Each male has his own special tone and when several are croaking at the same time it sounds like distant bell-ringing or like the noise made when one blows into the neck of an empty bottle.

This is essentially an aquatic animal which only exceptionally comes on land, usually at night. It jumps very well and when possible goes straight back into the water when disturbed. But it may also stand its ground and take up a defensive position, by bending the head and limbs up over the pliable back, thus showing part of the orange underside. It may stay in this position for several minutes.

During the breeding period in May–June the male grips the female with his fore-legs round her body in front of her hind-legs. The large eggs, 2–3 mm ($\frac{1}{8}$ in.) in diameter, are laid singly or in small grey-brown clumps of 25–50, with a total of 100–300 eggs. They hatch in about a week. The tadpoles may reach a length of 5 cm (2 in.) but the newly metamorphosed toads are only 1·5–2 cm ($\frac{2}{3}$–$\frac{3}{4}$ in.) long. The adults feed mainly on flying insects which fall into the water, and also on other insects and worms.

20 Parsley frog
Pelodytes punctatus

Identification: c. 4 cm (1$\frac{1}{2}$ in.). Body slender. Male has internal vocal sacs, and in the breeding season small dark brown horny pads on the inner side of the first and second fingers, on the upper and lower arm and on each side of the breast. The colour of the back varies from grey and brownish to olive with small green spots. Underside white. The skin can change colour very rapidly.

Distribution and habitat: Found in most of France, except the eastern part, north-western Italy and in the whole of the south and west of the Iberian Peninsula; it has also been recorded in Belgium. Always lives in the vicinity of ponds with dense vegetation, making a quick jump to the safety of the water when alarmed.

Habits: Essentially nocturnal, except during the breeding season. Swims very well and can jump a distance of 30–40 cm (12–15 in.). In the southern part of its range this species breeds in the spring from the end of February to May and again in the autumn in September–October. When mating the male grips the female with his forelegs just above her hind-legs, and usually remains in this position for a few hours. The eggs are laid in broad bands 6–8 cm (2–3 in.) long. The tadpoles are very small when they hatch,

but they grow fast and just before metamorphosis they may be 3 cm (1 in.) long, exceptionally 6·5 cm (2$\frac{1}{2}$ in.), and thus much longer than the adult frog. The croak, which is heard in the evening, is characteristic, reminding one of squeaking leather shoes.

21 Painted frog
Discoglossus pictus

Identification: 6–7 cm (2$\frac{1}{4}$–2$\frac{3}{4}$ in.). A sturdily built frog with relatively long legs. In the female the webs at the base of the tones are small, but in the male these are much better developed. He has only a rudimentary vocal sac and the croak is very weak. The skin is smooth or slightly warty. Coloration very variable. The ground colour of the back may be reddish, yellow-brown or greyish with dark brown markings, often with a pale border. Some individuals have a long pale yellow-brown dorsal stripe. Underside ivory-white, sometimes with brown spots.

Distribution and habitat: Found in southernmost France, Corsica, Sardinia, Sicily and Malta, and in the Iberian Peninsula except in the north-east. Lives mostly in bogs and ponds, but may also occur in hill streams and in brackish water.

Habits: Active both by day and night, often sitting in large numbers in shallow water near the shoreline with the head held up. During mating the male grips the female firmly around the loins. The female may lay several times during the year, the periods depending on the locality. The very small eggs are laid singly and only a few at a time. The tadpoles metamorphose when they are 1–2 months old.

22 Midwife toad
Alytes obstetricans

Identification: c. 5 cm (2 in.). Snout rounded and limbs very short. Back covered with small, round, smooth warts and there is also a longitudinal row of somewhat larger warts along each flank. The male has no vocal sacs and the webs include only about half the length of the toes. The colour of the back varies from ash-grey to pale brown with greenish or greyish markings. In the female the warts along the sides are often reddish. The underside is whitish-grey and finely granular. It is difficult to distinguish the sexes but the male has a slightly larger body and fore-limbs. Albinos occur occasionally and these may be white, pale red or pale yellow.

Distribution: Found throughout Spain and Portugal, in France except the south-eastern corner, in Belgium, southern Holland, the westernmost

districts of West Germany, particularly around the Rhine, and in northern Switzerland.

Habitat: Lives mainly in hilly country, sheltering under stones, in crevices in walls, in deserted mouse holes and mole runs, or in holes in the ground which it digs with its snout and fore-legs. In the Pyrenees it is even found at altitudes above 2,000 m (6,540 ft), where the ground is only free of snow for 3 months in the year.

Habits: Emerges from its shelter as darkness falls. It can run like a toad and also jump like a frog, but usually moves very slowly. The breeding season may extend over several months, from March to September, and during this period the female may lay 2, 3 or 4 times. The male seeks out the female on land and grips her round the loins with his fore-legs, but sometimes he slips forwards and may end up holding her round the neck. The female lays a band of relatively few large eggs. While sitting on the female's back, the male fertilizes the yellow eggs as they are laid and winds the egg band round his hind-legs. During the day he lies up in a sheltered place and if the weather has been dry he enters the water at night, thus preventing the eggs from drying out. After 2–3 weeks the male again goes down to the water with the eggs and when the tadpoles have gnawed their way out he tears off the empty egg band. During spring and summer one can find the tadpoles in all stages of development in ponds and also in areas that are temporarily flooded. Those tadpoles which hatch late spend the winter in the water, but the others generally metamorphose in 3–5 months. The newly hatched tadpoles are 1·5 cm ($\frac{2}{3}$ in.) long, and before metamorphosis they may be 5 cm (2 in.), exceptionally even 8–9 cm (3–3$\frac{1}{2}$ in.) long, whereas the newly metamorphosed Midwife toad only has a length of 2·5 cm ($\frac{7}{8}$ in.). Although they lack vocal sacs the males have a clear call, sometimes heard in the evening, which is reminiscent of good church bells. They feed on insects, larvae and worms.

23 Iberian midwife toad
Alytes cisternasii

Identification: c. 4 cm (1$\frac{1}{2}$ in.). Very similar to the preceding species but somewhat smaller. Back greyish or brownish with small dark markings. The warts on the eyelids and on the sides are whitish or orange.

Distribution and habitat: Found in the Iberian Peninsula, in Portugal from Oporto in the north to Lisbon in the south and eastwards in Spain to Estremadura and New Castile.

Habits: Lives in sandy places, where by day it burrows down, using the outer edge of the fore-feet. In other respects very similar to the preceding form, and here again it is the male that carries the eggs. The tadpoles of the two species are very similar to each other, and they can be found at any time of the year.

24 Common spadefoot
Pelobates fuscus

Identification: 6–8 cm (2$\frac{1}{4}$–3 in.), the female being the larger. Head broad and arched at the back. Skin smooth or with small flat warts. The male has no vocal sacs, but has a large, oval gland on the outer side of the upper arm. Toes webbed to the tips. Beneath the first toe there is a large yellow-brown, horny, spade-like growth, which has a sharp edge. The colour of the back varies considerably but is often pale brown in the male and pale grey in the female with olive or chestnut-brown markings mixed with small red spots. Underside pale grey, sometimes with grey-brown markings.

Distribution: Widespread in central and eastern Europe and into Asia, except in mountainous regions. In the west, this species extends to northern France, Switzerland and northern Italy, in the south to the Apennines and northern Balkans, in the north to northern Jutland, the extreme south of Sweden and Gotland and in the east to Leningrad and farther east.

Habitat: Essentially terrestrial, usually spending the day hidden in the ground. If disturbed when out during the day, the Common spadefoot digs itself backwards into loose soil or sand, using the powerful 'spades', on the feet, sometimes doing this in a few seconds. Prefers low-lying, sandy areas and only enters the water during the breeding season.

Habits: Emerges from its shelter in the evening, an hour or two after sunset. Produces a distinct smell of garlic, and so is sometimes known as the Garlic toad. It quite frequently utters a wailing screech, inflates the lungs and opens its mouth as though intending to bite. From the beginning of April to the end of May it goes down to marl-pits or to pools which dry out during the course of the summer. It only remains in the water for a few days. The eggs are laid in among water plants in bands of mucus about 18 in.

coal-black 'spades' on the hind-feet are large and sharp-edged. The skin is smooth or with very fine warts. Back yellowish or yellow-green, spotted or marbled with brown or grey-green markings.

Distribution and habitat: In the Iberian Peninsula and in western and southern France, particularly in coastal areas where it often lives on sand dunes.

Habits: Breeds at the end of March or April, a little later than the Common spadefoot, but its habits are otherwise very similar.

26 Natterjack
Bufo calamita

Identification: 6–8 cm (2¼–3 in.). Body short but not very thick, and legs shorter than in other anurans. Back covered with numerous small flat warts. The parotoid glands are rela-

long and the diameter of a finger. They hatch in about a week and the tadpoles, which are 6 mm (¼ in.) long, develop slowly, sometimes taking 4–5 months. They do, however, become very large, often 10–12 cm (4–4½ in.), and exceptionally 17–18 cm (6¾–7 in.). They are in fact the largest of all European tadpoles. By September they have gone on land after metamorphosing into toads which are only 3 cm (1 in.) long, but some of them may spend the winter in the water as tadpoles. The adults spend the winter buried in sandy soil. The diet consists of worms, snails, beetles and other insects which are hunted at night.

25 Southern spadefoot
Pelobates cultripes

Identification: 7–9 cm (2¾–3½ in.). Slightly larger than the Common spadefoot but otherwise very similar. Head large and very flat, and the shiny

tively small and the webs are short, only about half the length of the longest toes. Sexes practically the same size. Male with a strikingly large vocal sac which is particularly large during the breeding season. The colour of the back varies from grey or olive-green to brownish, marked with irregular grey or red-brown blotches. Along the back there is a narrow pale yellow stripe, which may sometimes be interrupted but is rarely absent. Underside pale grey in front with black spots and greyish-black behind with white dots.

Distribution: Essentially a west and central European form, with its eastern limits running from the Baltic States and Gulf of Finland southwards along the eastern borders of Poland and then south-westwards to the Alps. In the west the range extends to England and south-western Ireland and in the north to Denmark and the extreme south of Sweden.

Habitat: Often occurs in sandy places such as on dunes and tolerates drought and a relatively high salinity. In mountainous areas it can be found at altitudes of more than 1,000 m (3,270 ft). By day it shelters under stones, in mouse runs or most frequently in holes that it has itself dug. It usually digs down into the sand backwards, using the horny tips of the toes. When it has dug a certain way down it will turn round and go on digging with the snout and fore-legs, scraping the sand away with the hind-legs.

Habits: Emerges from hibernation at the beginning of April and enters the water, often pools with brackish water, where breeding goes on until the end of May and sometimes until later in the summer. The female lays a band of

relatively few eggs arranged in one or two rows, which is twisted around reeds and water plants. Egg-laying takes place at night and lasts only a few hours. About 6–7 weeks after the eggs have been laid the tadpoles metamorphose into toads which leave the water when scarcely more than 1 cm ($\frac{1}{3}$ in.) long. Thanks to its large vocal sac the male has the loudest voice of all European toads, giving a trilling growl from shortly after sunset. Usually a number will croak together, and they can be heard over long distances. Unlike most other frogs and toads the Natterjack runs rather like a mouse, stopping every foot or two and resting, hence its alternative name of Running toad. It can jump, but clumsily and only for very short distances. Diet consists of insects and larvae which are caught at night.

27 Green toad
Bufo viridis

Identification: 6–9 cm ($2\frac{1}{4}$–$3\frac{1}{2}$ in.), exceptionally longer. Body very short, the female being a little larger than the male. The back has numerous warts, but these are not very large, and the parotoid glands are flat and kidney-shaped. Male with a relatively small vocal sac which is divided into two internally. Webs only weakly developed. Coloration varies considerably, the back being pale grey or olive-green with irregular grass-green markings with black edges and often with red warts, particularly on the flanks. When sitting in the shade this toad may be an almost uniform dark greenish-brown, but if it comes out into the light the colouring will change, in about 10 minutes, to the more charac-

teristic pale grey with grass-green blotches.

Distribution: Essentially an east European form, with its western limits running roughly from Bremen southwards along the Rhine and on to about Genoa. To the north it occurs in Denmark and southern Sweden and eastwards to the Baltic States.

Habitat: Lives in coastal and sandy areas and in dikes, sheltering by day under stones or in crevices. In sandy places the Green toad can dig holes and burrows, using its hind-legs. Reaches altitudes of 2,000 m (6,540 ft) or even more in mountainous regions.

Habits: Emerges from hibernation in April, and breeding lasts from then until the beginning of June, sometimes going on even later in the summer. This species tolerates brackish and salt water. The voice of the male is a long,

trilling whistle which can be heard over long distances. Jumps and swims better than the other toads. The female lays some 7,000–12,000 very small eggs in a strand that is 3–4 m (10–13 ft) long. Metamorphosis is completed in August, and by September the old and young toads are moving into their winter quarters.

28 Common European toad
Bufo bufo

Identification: 8–12 cm (3–4½ in.). The size increases from north to south in Europe, and females over 18–20 cm (7–8 in.) are found in southern Italy and in Sicily. This is, therefore, the largest European anuran. Body broad and plump, head short and rounded and snout short and truncated. Parotoid glands large, prominent and the shape of a half moon. Fore-legs powerful, especially in the male. Toes short and united by a web. The male

has no vocal sac. Skin warty, and often some of the warts on the back and limbs are horny. The skin is thick, and on the back it may contain calcareous deposits. The upperside may be brownish-black, grey-brown, red-brown or olive, the colour varying according to sex, age, season and locality. Young individuals are usually yellow-brown or coppery-red, roughly like the adult females but older males tend to be more greyish or rather olive in colour. The underside is nearly always greyish.

Distribution: This is certainly the most widely distributed of all European anurans. In fact it occurs almost everywhere in Europe, as far north as 65° N, but is absent from Iceland, Ireland, Corsica, Sardinia, the Balearics and a number of small islands. In the Alps it extends up to altitudes of 2,000–2,500 m (6,540–8,200 ft).

Habitat: Spends the day in woods, gardens and fields, sheltering under stones, tree roots or in among dense vegetation. By night and also in dark, rainy weather it crawls or jumps around searching for food. Also known to enter cellars.

Habits: Emerges from hibernation early in spring, and sometimes a couple of days of sunshine can entice it out even earlier. In spring, mainly in April, this species spends effectively all its time in the water, the pairs remaining united both by night and day; at this time the croaking of the males goes on almost continuously. It is not known how the toads find the breeding-places, but probably a sense of smell plays a decisive part, for they are able to recognize the smell of certain species of algae in their own ponds. If a male and female meet on the way to the breeding pond he will climb on her back, holding on with his fore-legs in her armpits, while she continues on her way. The females are always in a minority, and when the pair reach the pond there will be fighting as several males throw themselves on her. Sometimes a whole cluster of entangled toads can be seen splashing around in the water. The female lays 5,000–7,000 small black eggs, arranged in 2–4 rows in a band about 3–5 m (9·8–16·4 ft); usually she lays the eggs in batches at varying intervals. The male fertilizes the eggs after they have been laid, and uses his hind-limbs to help pull out the egg band, which becomes wound around water plants. The pairs are often united for about 10 days, and sometimes for 3–4 weeks. The eggs hatch in 8–10 days and the tadpoles creep out of the mucus and for their first few days hang from the remains of the egg band, attached by their adhesive organs. Although the adult is relatively large, the tadpole is one of the smallest, measuring only 4–6 mm ($\frac{1}{4}$ in.) on hatching, and not more than 2–3 cm ($\frac{3}{4}$–1 in.) at metamorphosis. These tadpoles are black with a short tail, and they move around in small shoals. It is only at the end of metamorphosis that they become brown. Metamorphosis takes place in June–July and the young toads—only 1 cm ($\frac{1}{3}$ in.) long—go on land. They do not become sexually mature until they are 4 years old. Toads live to a great age, certainly for as long as 40 years. They do not tolerate drought at all well but can go without food for months at a time. They hibernate on land in places similar to those in which they shelter during the day.

Diet consists mainly of worms, slugs, wood-lice, insects and larvae, but sometimes they take the young of grass snakes or slowworms and also small frogs. When hunting they creep up to the prey very carefully, look fixedly at it with the head held askew and take aim; at this time one can often seen the toes twitching. All of a sudden the sticky tongue is shot out to catch the prey, but quite often the toad misses. When one approaches a toad it will either press itself down against the ground or hop away in a clumsy manner. If it meets something unexpected it will take up a defence position, by inflating the lungs and blowing itself out, at the same extending the legs, and raising its back. This can happen if a snake glides in under the bush where it is sitting or comes gliding in between its fore-legs, or if a hedgehog starts to attack. It will hold this position for a few seconds, then sink back into the resting position before going through the whole performance again. Toads are often killed by cats, polecats and other predators, but they are not eaten on account of the bitter, possibly lethal secretion from the numerous glands, particularly the parotoid glands. Snakes, however, do eat them. Toads are beneficial amphibians as they consume large numbers of snails and injurious insects in nurseries and private gardens.

29 Field frog
Rana arvalis

Identification: 6–8 cm ($2\frac{1}{4}$–3 in.), the female being the larger of the two. Body very slender, head as long as it is broad and snout protruding forwards from the mouth like that of a shark. The diameter of the tympanic mem-

brane is less than that of the eye. Limbs longer than in the Common frog, but the hind-legs relatively short and the nuptial pads are hard and prominent. During the breeding season the male's web increases in size, and in the female it has a deeply concave edge. Skin usually smooth, but in the mating season it may be somewhat granular and thickened. The rearmost part of the thighs and the belly are slightly rough. The male has internal vocal sacs and powerful fore-legs and during the breeding season his thumb pad and an area on the side of the thumb are covered with small, fine, dark horny outgrowths. The colour of the back varies considerably and may be pale yellowish, brownish, greyish or reddish, often with dark spots or stripes. Sometimes it may be completely covered with small black spots. Underside uniformly white or yellowish without spots. During the spring one may find lavender-blue males, a skin colour caused by the lymph in glands just below the skin but this colour is only retained for a short time.

Distribution: Northern and central Europe. In the north it extends to southern Norway and Sweden and eastwards far into Russia. In the south it reaches the northern Balkans, from where its western limits runs roughly from northern Yugoslavia, north of the Alps, along the Rhine to the Netherlands.

Habitat: Almost exclusively in low-lying country, although sometimes in the hills. Lives mainly in bogs, fields and meadows in the vicinity of ponds and lakes and is active almost the whole day. Enters hibernation in November and it is said that the males spend the winter in the mud of lakes, the females on land.

Habits: Usually emerges at the beginning of March, sometimes 1 or 2 weeks later than the Common frog. The croak of the male is like a hissing gurgle and has been likened to the sound of air bubbles escaping from an empty bottle held under water. The female lays 1,000–2,000 relatively small eggs which sink to the bottom where they develop. The rate of development of the tadpoles is largely dependent upon the weather. They reach a length of 3–4 cm (1–1½ in.), but the newly metamorphosed young frogs are only 1–2 cm (⅓–¾ in.). They become sexually mature when about 3 years old. Diet consists of insects, worms and other invertebrates. Their enemies include foxes, badgers, otters, polecats, storks, herons, buzzards and snakes.

30 Common frog
Rana temporaria

Identification: 7–9 cm (2¾–3½ in.); the female may be even larger. This species is a little larger than the Field frog and it is often difficult to distinguish them. However, the Common frog has a somewhat shorter, more rounded snout and small, round, soft nuptial pads on the first fingers. Body not so slender as in the Field frog and the head is broader than it is long. The diameter of the prominent tympanic membrane is only three-quarters that of the eye. Limbs shorter than those of the Field frog, nevertheless a Common frog can jump a distance 6–7 times its own length. The webs are well developed in both sexes but they never extend farther than the penultimate joint of the longest toe. Skin smooth or with small, flat, irregular warts. The rearmost parts of the limbs are somewhat rough. The males have two internal vocal sacs which are inflated during croaking and appear as swellings on each side of the throat. Fore-limbs very powerful, particularly during the spring, at which time the thumb pads and the sides of the thumbs are covered with small, black horny outgrowths. The skin of the females is more granular. This species is so variable in size, colour and pattern that one literally never finds two specimens exactly the same, and this accounts for the long series of more or less well-established varieties. The ground colour of the back varies from yellow through brown, red or grey to almost completely black. The males tend to be darker than the females. Although the patterns are so variable there is nearly always a dark stripe running along the inner side of each fore-leg, an inverted

V between the shoulders and temporal markings which end behind in a point. The hind-legs usually have dark transverse bands. Underside whitish or pale yellow with grey marbling. During the breeding season there may be whole populations in which the males have blue throats or are even blue over the whole body, as in the Field frog, but this coloration is quickly lost.

Distribution: This is the most widespread of all the European frogs, being found in all parts of the Continent except the Mediterranean countries. It is the only European frog to extend right up to the North Cape in Norway. The southern limit runs from southwest France through northern Italy to Yugoslavia and Bulgaria; there is an isolated population in the north-west corner of the Iberian Peninsula. It is also found in Asia eastwards to Japan.

Habitat: Active throughout most of the day, in fields and meadows, wood-

land and gardens, where there are damp, shady places, and it extends up to altitudes of over 2,500 m (8,200 ft). It usually hibernates in the mud at the bottom of lakes, or sometimes on land.

Habits: Emerges from hibernation at the beginning of March, or sometimes even earlier, and egg-laying goes on from March to April. The male clasps the female in such a way that his fingers meet beneath her breast, and the males, which are in the majority, may be just as aggressive in clasping as those of the Common toad. Like many other species of frog and toad the males will try to mate with any other moving object of a similar size and consistency. But if a male catches another male and tries to clasp it, the latter emits a distinct call which causes the former to release his grasp immediately. When a male tries to clasp a female who is not ready for egg-laying, or who has already laid her eggs, she bends her body and tries to kick him off and this is usually sufficient to make him give up the attempt. Ordinarily this frog makes no sound, but a purring croak can be heard during the breeding season. At a distance this may sound like a faraway railway train if all the frogs, including the females, croak at the same time. They usually croak under water. The female lays her eggs relatively quickly and she may produce and have fertilized up to 1,000 eggs in an hour; a complete spawning may total 3,000–4,000 eggs. The egg clumps sink to the bottom but after the albumen has swollen they rise and remain floating at the surface. The yolk is 2 mm ($\frac{1}{16}$ in.) in diameter, the albumen about 10 mm ($\frac{3}{8}$ in.). In some places the eggs may be present in hundreds of thousands, and so closely packed that they form a huge mass of jelly. Since egg-laying takes place so early in the year, the development of the eggs may take a relatively long time, and it may be 3 weeks before they hatch into tadpoles that are 6–8 mm ($\frac{1}{4}$–$\frac{5}{16}$ in.) long. The subsequent development is usually complete 3 months after the eggs were laid. Exceptionally the tadpoles may spend the winter in the water. They can reach a length of 4·5 cm (1$\frac{3}{4}$ in.), whereas the newly metamorphosed frogs are only 1·0–1·5 cm ($\frac{1}{3}$–$\frac{2}{3}$ in.). Sometimes they go on land in such enormous numbers that the ground may be literally covered with them, and indeed millions may leave a single lake. These young frogs spread out into the surrounding areas and do not return to the lake until they are sexually mature at an age of 2–4 years. Common frogs feed on insects, worms and other invertebrates.

31 Italian agile frog
Rana latastei

Identification: 4–6 cm (1$\frac{1}{2}$–2$\frac{1}{4}$ in.), exceptionally 7 cm (2$\frac{3}{4}$ in.). Skin smooth or with a few small warts. Webs very well developed in the male. No vocal sacs. The colour of the back varies from grey to brick-red, being either uniformly coloured or faintly spotted.

Distribution and habitat: Southern Switzerland and northern and central Italy, living in open deciduous woodland with small streams and rivers. Mostly found in low-lying country and does not extend to altitudes above about 800 m (2,600 ft). This and the following 3 species are all closely related and very similar to one another.

Habits: In the breeding season the Italian agile frog leaves the woodland and moves into ponds in the meadows. It lays eggs a little later than the next species, but otherwise its habits are similar.

32 Agile frog
Rana dalmatina

Identification: 5–7 cm (2–2$\frac{3}{4}$ in.), the females being larger than the males. Body slender, head flat and as long as it is broad, snout pointed and protruding quite a bit in front of the mouth. The tympanic membrane is close behind the eye, and the same size as the latter, or larger. Fore-limbs relatively short, but hind-limbs strikingly long, and this applies particularly to the shins. The thumb pads are hard, oval and prominent. The skin is tighter than in most other frogs and is smooth, except on the hindmost part of the belly where it is slightly granular. The male has no vocal sacs and has shorter fore-legs than the female. The sexes only show external differences during the breeding season when the male's thumb pads are dark with small horny outgrowths and his webs are strongly developed. The brown colour of the back may grade into pale grey or muddy yellow. This frog may present a rather transparent appearance, and a dark inverted V behind the head is often the only colour pattern on the back. The hind-legs have regular, dark transverse bands, and the underside is uniformly white or yellowish.

Distribution: France, Switzerland, Germany, Italy, Austria, the Balkans and eastwards into Asia Minor and the Caucasus. There are also isolated

them. The eggs at first sink to the bottom but when the albumen has swollen they rise to the surface. The tadpoles develop very slowly and metamorphose after 3–4 months. This species is well named, for although relatively small it can jump a distance of 2 m (6–7 ft) and to a height of 2–3 m (6–10 ft).

33 Spanish frog
Rana iberica

Identification: c. 6 cm (2¼ in.). Skin completely smooth or with very small warts. Male with no vocal sacs. The colour of the back varies from yellow-brown through grey to reddish, with yellowish or dark brown spots.

Distribution and habitat: Restricted to the Iberian Peninsula where it is found in northern Portugal southwards to Lisbon, in north-west Spain and in the Pyrenees. Lives mostly in the hills in the vicinity of springs and streams and at altitudes up to 1,500 m (4,900 ft).

Habits: Very similar to those of the preceding species.

34 Greek frog
Rana graeca

Identification: 6–7 cm (2¼–2¾ in.). Skin smooth or with fine warts. The male has very powerful legs and an external vocal sac. Back grey or brownish, usually with scattered dark markings, but it may also be reddish, yellowish or olive-green.

Distribution and habitat: Found in the Balkans from Bosnia, central Macedonia and southern Bulgaria to southernmost Greece, and also in the Apennines in Italy. Lives close to

populations on the larger Danish islands, in the coastal districts of south-west Sweden, and on the neighbouring island of Öland.

Habitat: Open deciduous woodland, sunny banks and meadows. The females possibly hibernate on land, the males at the bottom of small lakes.

Habits: Hibernation is short and even when there is still ice on the water the males can be seen beneath it, As soon as the ice has melted the frogs are ready to breed, but the females, which are very much in the minority, usually come out somewhat later than the males. The croak of the males in spring is similar to that of a tree-frog, but is much weaker, as they have no vocal sacs. The Agile frog is not particular about its place of spawning, and mating probably only lasts for a few hours during the night. The parents leave the water as soon as the female has laid the eggs and the male has fertilized springs and streams and extends up the mountains to altitudes of 2,000 m (6,540 ft).

Habits: Similar to those of the Agile frog and the Italian agile frog, but as it mostly lives up in the mountains it usually breeds somewhat later than its close relatives.

35 Edible frog
Rana esculenta

Identification: 7–10 cm (2¾–4 in.), the female being the larger and exceptionally reaching a length of 11–12 cm (4¼–4½ in.). Head flat and triangular with a short, pointed snout. Fore-legs short, and more powerful in the male. Hind-legs relatively short in this species, whereas the toes are very long, with well-developed webs. Skin smooth or somewhat gritty to the touch, due to tiny warts. The male's vocal sacs lie hidden in a groove

behind the mouth and appear externally as a pair of bladders almost the size of a hazelnut when the frog croaks. During the mating season the male's thumb has a swollen rough pad which is not so thick and dark as in most other anurans. Upperside mostly green, particularly in the male, but may also be brownish or bluish. The tip of the snout and the edge of the upper jaw are nearly always black. Back more or less blotched with black, and along the mid-line there is a pale green stripe which may sometimes be almost yellow. The limbs also have dark markings which on the hind-legs may form transverse bands. Underside of body white or greyish, frequently with dark markings.

Distribution: Widespread in most parts of Europe northwards to Denmark, south Sweden and south Finland. Also extends into parts of northern Africa and in Asia eastwards to Afghanistan. It has been repeatedly introduced into Britain but has usually disappeared after a time; there are still colonies to the south-west of London and in Kent.

Habitat: Essentially an aquatic frog. After egg-laying it remains in the water or lives along the banks of a lake, always ready to jump into the water. In autumn it digs down into the bottom of the lake where it hibernates.

Habits: In spring the juveniles emerge before the adults. About the middle of May, but earlier in the southern part of the range they can be seen along the banks of lakes and at the end of the month the croaking of the males can be heard day and night, for miles around. Each time a male croaks the vocal sacs inflate like a pair of white balloons. In contrast to other frogs the Edible frog does not begin to spawn immediately, but spends a certain amount of time feeding. The eggs are laid from the end of May to the beginning of July, or even later, after a short period of clasping which probably takes place at night, for they are seldom seen pairing. The male clasps the female under the arms so that his hands almost meet under her breast. The eggs are laid in several clumps, often with over 1,000 whitish-grey eggs in each, attached to water plants near the surface. Egg-laying goes on for at least a month and the eggs from a single female will be fertilized by several different males. The eggs are 1·5 mm ($\frac{1}{16}$ in.) in diameter and a single female may lay a total of 5,000–10,000 in a season. The newly hatched tadpoles are at first brownish-black but the upperside soon becomes olive with brown spots and the underside mother-of-pearl with yellowish markings. Development usually lasts for 3 months, and in the course of the last week the gill-breathing, vegetarian tadpoles with a tail for swimming changes (metamorphoses) into a lung-breathing, tailless predatory frog which goes on land and moves about with the help of its long hind-legs which are adapted for jumping. It is, however, not uncommon for some of the tadpoles to spend the winter as such. The young frogs are 2·5 cm ($\frac{7}{8}$ in.) long and they often go into hibernation immediately after metamorphosis, with scarcely any chance to feed. The tadpoles are among the largest in Europe, being normally 8–9 cm (3–3$\frac{1}{2}$ in.) long, exceptionally 10 cm (4 in.) or even longer. The Edible frog is very voracious and will eat anything moving that it can get into its

mouth, and particularly insects, larvae, snails and worms, but sometimes also small lizards and snakes.

36 Marsh frog
Rana ridibunda

Identification: 9–15 cm (3$\frac{1}{2}$–6 in.), exceptionally up to 17 cm (6$\frac{3}{4}$ in.). Very similar to the Edible frog in coloration and habits, so specimens of the same size may be difficult to distinguish from one another. This is why the Marsh frog was for a long time regarded as a variety of the Edible frog. The Marsh frog has longer hindlimbs and the warts on the skin are larger and more numerous. The back is often olive-green or brownish. The sides of the body and the rear of the thighs are green or greyish-white, whereas in the Edible frog they are yellow. The dark markings on the back are larger, without such sharp edges and the Marsh frog's vocal sacs are greyish, whereas those of the Edible frog are whitish.

Distribution: Mainly an East European species with its western limits along the Rhine, and known mainly from Germany, Poland and western Russia. Introduced into the Romney Marshes area in Kent in 1935 and now covers an area of about 30 square miles.

Habitat: Usually lives in the open water of lakes, far from the shore, particularly in places with dense vegetation, and also occurs in running water.

Habits: Breeds in April–May, about 14 days earlier than the Edible frog. The croak is very different from that of the Edible frog, being louder and less continuous, and as the scientific name suggests it is somewhat like human laughter. It often starts to croak when rainy weather is approaching. A very voracious species which, in addition to insects, larvae, worms and snails, also eats fish, other anurans, lizards, snakes and mice.

37 European tree-frog
Hyla arborea

Identification: 3–5 cm (1–2 in.). One of the smallest European frogs. Body slender, and very narrow just in front of the hind-legs; head broad with a blunt snout. The toes are joined by a deeply indented web, and the tips of all the fingers and toes have small round adhesive discs. Skin completely smooth on the back, but on the underside with small granules. Upperside usually bright green, but the colora-tion is affected by age, locality, lighting, sloughing periods and mating, and it may be green, yellow, grey, blue, brown or even black. Not all individuals are equally adept at changing colour. Every fortnight, immediately after sloughing, the coloration may be grey-blue, possibly with marbling. The coloration of the back is separated from that of the belly by a narrow stripe, black outside and whitish-yellow inside. This stripe begins at the nostrils and continues down along the body to the groin; in this area it forms a branch, known as the hip sling, and often continues out on to the hind-legs. Underside whitish, fingers and toes flesh-coloured. Throat dark brown in the adult male, a colour due to the underlying vocal sac. The male lacks the thickened areas of horny skin used by most other anurans for holding the female firmly; he grips her in the arm-pits.

Distribution: Southern and central Europe, northwards to about 56° N, but not in Britain and in Sweden only in a few localities in the extreme south.

Habitat: Lives in the water only during the period of egg-laying, and at other times is mostly seen in trees, bushes and among reeds. In good weather these frogs sit on the upper-sides of leaves, but will move round on to the undersides when the sun is too hot or when it begins to rain. They hibernate under the roots of trees or in holes in the ground.

Habits: Breeds in clear, standing waters with dense vegetation. Mating usually takes place in May–June and only lasts a couple of days. The female lays 800–1,000 eggs in a few yellow-ish-brown clumps which sink to the bottom. After 10–12 days the eggs hatch into tadpoles which are scarcely 5 mm ($\frac{1}{6}$ in.) long. These are at first whitish-yellow but soon become darker with golden or iridescent mother-of-pearl markings. Later still they become yellow-green on the upper-side. Development is completed some three months after egg-laying, and the newly metamorphosed tree-frogs are not more than 1·5 cm ($\frac{2}{3}$ in.) long.

They are not fully grown and sexually mature until the fourth year.

The rapid, very loud croak of these attractive frogs can be heard at a considerable distance, particularly during the spring. When fully inflated the vocal sac may be twice as large as the head. Tree-frogs can also be heard croaking in the summer during rainy or thundery weather. They feed on flies and other small insects which they sometimes jump after and catch in mid-air.

Chelonians

The chelonians (tortoises, terrapins and turtles) form a relatively small group of reptiles, containing about 250 species, of which only 11 occur in Europe or off its coasts. The body is encased in an armour, the shell, consisting of a dorsal carapace and a ventral plastron. The carapace and plastron have an internal structure of dermal bony plates which are covered externally with large epidermal scales known as laminae or shields, the boundaries of which do not correspond with those of the underlying dermal bones. The other parts of the body are covered by horny scales or small horny shields. The toothless jaws have horny sheaths with sharp edges. Unlike other higher vertebrates chelonians cannot breathe by expanding the chest, and have evolved their own method. Air is inhaled when two flank muscles contract, increasing the volume of the body cavity around the lungs, and it is exhaled when two pairs of ventral muscles contract and press the internal organs against the lungs. Some aquatic chelonians can absorb oxygen from water that is pumped in and out of the throat, and others pump water in and out of two sacs that open into the cloaca. However, air-breathing is much more important than either of these two methods. With the exception of the marine turtles, all the European chelonians spend part of the year in hibernation. The males, which are smaller than the females, have a concave plastron, which more or less keys with the convex carapace of the female during mating. The female digs a pit for the eggs which usually have white calcareous shells. She covers the eggs with sand or vegetation, and they hatch 2 or more months after having been laid, sometimes a whole year later. Some tortoises live for upwards of 100 years, possibly more, but 50 years can be regarded as a good age for most of them.

38 European pond-tortoise
Emys orbicularis

Identification: 12–25 cm (4½–10 in.), in the southern part of the range up to 35 cm (14 in.). These figures refer to the length of the carapace, and the same applies for the other chelonian species. The carapace is oblong, slightly domed, broadest at the rear and united at the sides to the plastron. The flat, or in the male slightly concave plastron is divided into two by a transverse joint, giving a smaller front part and a larger rear part and these are slightly mov-

able. The head, limbs and tail can be withdrawn under the carapace. The flat feet are webbed and the toes have claws. The tail is about half as long as the carapace in the female and about two-thirds in the adult male. The black or dark brown carapace nearly always has yellow spots or stripes, their number and arrangement being very variable. The plastron tends to be uniformly yellow or yellow with a blackish-brown pattern.

Distribution: Scattered over most of southern and central Europe, north-wards to the Baltic Sea, but absent from the Netherlands and most of West Germany. Also found in western Asia and north-west Africa. Absent from Britain.

Habitat: Mostly in small lakes, bogs and swamps with rich vegetation. Although it likes to lie out on the banks in the sun this is a shy animal which moves into the water at the slightest disturbance.

Habits: In the autumn it digs down into mud or soft ground near the water and emerges again in March–April–May, depending upon the climate and latitude. Mating usually takes place from the middle of May to the beginning of June. Before laying her eggs the female bores into the ground with her tail and then turns herself round, thus forming a conical hole, which she then deepens by digging with the hind-limbs. In this hole she lays 5–15 white, oval eggs with tough, parchment-like shells which harden soon after. When the eggs have been laid she covers them with soil, pats this down with the plastron and then leaves the warmth of the sun to incubate them. They usually hatch in August–September but in northern latitudes the young overwinter in the eggshells and emerge in the following spring. When ready to hatch the young one first makes a hole with the claws of one fore-leg, then another hole with the other leg and finally tears open the part between the two holes with the egg-tooth on the snout. During the evening and at night pond tortoises hunt in the water for worms, snails, bivalves, small fish, frogs and newts. If they find food on land they take it into the water to eat it. They will either swallow the prey whole or cut it into pieces with the sharp horny jaws.

39 Spanish terrapin
Clemmys caspica leprosa

Identification: 18–20 cm (7–8 in.), in the southern part of the range up to 30 cm (12 in.). The principal external features are approximately the same as in the preceding species, but the plastron is not divided into two by a transverse joint. The carapace is usually olive or red-brown with a pattern of irregular, mostly yellow lines which disappears with age. The neck is marked with yellow or orange longitudinal stripes.

Distribution: There are 3 subspecies of *Clemmys caspica*. The first, described here, occurs in the Iberian Peninsula

(except the most north-easterly part) and in north-west Africa. The second is found in south Yugoslavia, Albania, Bulgaria, Greece, Turkey, Syria and Israel, and the third in the Caucasus and Iran.

Habitat: Lives in streams and rivers and also in ponds, lakes and brackish-water lagoons, but can also survive for a time on land, often quite far from water.

Habits: Very similar in general habits, diet and breeding to the European pond tortoise.

40 Hermann's tortoise
Testudo hermanni

Identification: 20–25 cm (8–10 in.). As in the other land tortoises the carapace has a high dome and the flat plastron is slightly concave in the males. The legs are almost cylindrical and the fingers and toes are fused into club-feet with short claws. The horny plates of the carapace are black, yellow and brown but the coloration is quite variable. The plastron is paler with 2 longitudinal rows of dark spots. The plate at the rear of the carapace is known as the supracaudal and in this species it is double. The short tail, which is shortest in the female, ends in a hard horny spur.

Distribution: Only in southern Europe, where it is common in north-east Spain, south France, Italy (except the north-eastern part) and throughout the Balkans. Also in the Balearics, Corsica, Sardinia and Sicily.

Habitat: Lives mostly in open, undulating country where it is dry and warm, but seldom extends to altitudes above 700 m (2,300 ft). In spite of its

November. Diet consists mainly of fruits, grasses and leaves, but they also eat worms, larvae and snails. They can certainly live to an age of 50 years.

41 Greek tortoise
Testudo graeca

Identification: 20–30 cm (8–12 in.). Very similar to Hermann's tortoise, but may grow a little larger. It is distinguished by having the supracaudal plate undivided and by lacking the hard, horny spur at the end of the tail. It also has a prominent horny tubercle on the back of each thigh. The distribution of colour varies as in Hermann's tortoise, but the plastron usually has black along its middle line.

Distribution: Found in two distinct areas of Europe: one, in south-eastern Spain and on Ibiza, and the other in

preference for dry places it drinks a lot and likes to bathe.

Habits: Emerges from hibernation in February–March–April, depending upon climate and latitude. Mating takes place early in the spring. The male pursues the female and pushes his head in under her carapace and knocks against her. He also tries to make her receptive by biting her legs. In May–June the female digs a pit with her hind-legs and lays 5–10 almost spherical eggs with hard, calcareous shells about the size of a hazel nut. She may lay again after an interval of a week. The warmth of the sun incubates the eggs, which hatch in September. During the summer these tortoises dig down into the ground at night and may do the same during the hottest part of the day. The temperature for normal activity is between 16°C and 32°C (60°F and 89°F). They usually go into hibernation in October–

the Balkans, particularly to the east of the Danube. Also widespread in north African and south-west Asia.

Habitat: Lives mainly in very dry regions, and also in hill forests up to altitudes of 1,100 m (3,600 ft).

Habits: Almost exactly the same as those of Hermann's tortoise.

42 Loggerhead turtle
Caretta caretta

Identification: 60–90 cm (23–35 in.), exceptionally over 100 cm (40 in.). Carapace oval to heart-shaped and only slightly domed, head large and broad. The carapace is almost chestnut-brown, the plastron yellowish. Like the following 3 species the Loggerhead turtle spends its whole life in the sea, except when laying eggs. The large fore-limbs form flippers, with all the joints immovable, except at the shoulder, and with the fingers enclosed in a common skin. The hind-limbs are flipper-like, with claws on 2 toes at the most. Tail short. The head cannot be withdrawn beneath the carapace.

Distribution and habitat: Lives in tropical and subtropical seas, is very common in the Mediterranean and is also found in the Black Sea and Bay of Biscay. Occurs both in the open sea and in rivers. Occasional specimens are recorded from time to time off the coasts of France, Belgium, Holland and Britain.

Habits: Feeds mainly on animal food, particularly crustaceans and molluscs. Breeding habits very similar to those of the Green turtle (No. 44). The horny plates have no commercial value and the flesh is not particularly

well-flavoured, but the eggs are taken by man.

43 Hawksbill turtle
Eretmochelys imbricata

Identification: 45–60 cm (17–23 in.), exceptionally 90 cm (35 in.). The carapace is oval to heart-shaped with brown plates usually marbled with yellow. In young specimens the plates overlap like the tiles on a roof, but in adults they lie almost edge to edge. The plastron is yellowish and the upper jaw ends in a hooked beak, hence the popular name.

Distribution and habitat: Along the coasts of the Atlantic Ocean and Mediterranean Sea, occasionally straying into the North Sea and to the coasts of the British Isles. Also widespread in the Pacific and Indian Oceans.

Habits: Breeds in the same way as the Green turtle (No. 44). Feeds on marine plants and animals, particularly on fish.

This is the chelonian which yields true tortoiseshell which is used for combs, boxes and handbags. Each specimen yields 2–3 kg (4½–6½ lb). In Cuba they catch Hawksbill turtles by attaching a line to the tail of a sucker-fish which is then allowed to fix itself to a turtle.

44 Green turtle
Chelonia mydas

Identification: 90–115 cm (35–45 in.); very large specimens may be 150 cm (60 in.) long and weigh 450 kg (about 1,000 lb), but most of those caught weigh 30–70 kg (65–150 lb). The carapace is oval or heart-shaped and brown or olive in colour with yellowish or brownish marbling. The plastron is yellow.

Distribution and habitat: Widespread in the Atlantic, Pacific and Indian Oceans, and also in the Mediterranean Sea. In some years this species drifts on to the coasts of Belgium and the Netherlands.

Habits: An excellent swimmer which is often seen far out at sea, although it is primarily a coastal animal which feeds mainly on various seaweeds. The young also eat crustaceans, bivalve molluscs and squid. It sometimes goes on land to lie in the sun and to sleep, but it can also sleep at the surface of the sea. After mating the females crawl up on land to lay their eggs at a place above high water; this usually happens during the night. They dig a large hole in the sand using the hind-limbs alternately, like a pair of shovels. Each female usually lays 2–5 times in the season, each clutch containing 70–200 eggs. After laying she covers up the eggs with sand, carefully smoothing the area with her flippers, and then returns to the sea. The eggs are incubated by the warmth of the sand, heated by the sun, and they usually hatch in 6–9 weeks. The baby turtles crawl out of the sand and immediately scuttle down to the sea, but even on this short journey many of them fall prey to frigate birds and other predators. Those that survive swim out to sea where they start to feed and grow.

This is the chelonian from which turtle soup is made, and the adults and eggs also yield an oil. At one time Green turtles were very abundant in all tropical and subtropical seas, but now there are large areas where their existence is seriously threatened, largely owing to the collection of the eggs by man.

45 Leathery turtle
Dermochelys coriacea

Identification: 150–270 cm (60–106 in.). This, the world's largest chelonian, has an average weight of 300-400 kg (660–880 lb), but there are records of specimens weighing over 600 kg (1,300 lb) and it is, in fact, the heaviest of all living reptiles. It has a superficial resemblance to the other marine turtles but instead of large scales it has a large number of small polygonal plates arranged like a mosaic in the thick, leathery skin. The carapace is flat and roughly heart-shaped, with 7 raised longitudinal ridges or keels; the plastron has 5 similar ridges. The head is slightly pointed and the tail very short. The limbs are very long, particularly the fore-limbs, and they have no claws. Young Leathery turtles are dark brown or black with pale spots or longitudinal stripes. The carapace becomes a paler brown with age, but the limbs remain dark.

Distribution and habitat: Widespread in all tropical seas, but nowhere in large numbers. It is frequently seen around southern South America, Africa, Australia, Japan, British Columbia and Nova Scotia. Dead or dying specimens are occasionally driven ashore on the coasts of France, Holland and western Germany and more frequently on the coasts of the British Isles. They also appear from time to time in the Mediterranean.

Habits: On moonlit nights the females crawl up on to deserted beaches, using their flippers to propel themselves laboriously along the sand. Each female digs a hole about 3 ft deep, and then lays a clutch of 60–150 eggs. She will then cover the eggs by spreading sand over them with the hind-limbs and then crawl down to the edge of the sea where she rests a while before swimming away. She may lay several such clutches during the year. The spherical, white eggs are about 5 cm (2 in.) in diameter and they hatch in about 2 months. The newly hatched young, which are 6 cm ($2\frac{1}{4}$ in.) long, crawl down to the sea where they can swim and dive immediately. Leathery turtles feed on fish, molluscs, crustaceans and jellyfish.

Lizards

The lizards comprise over 3,000 species, of which only 63 are found in Europe. This is the most variable of the reptile groups. In the most typical lizards the body is elongated with two pairs of 5-toed limbs and a long tail, but other lizards are snake-like without limbs or only with rudiments. The teeth grow either on the ridge of the jaw or on its lateral surface. Lizards of the Viviparous lizard type (lacertids) have the upper arms and thighs directed outwards from the sides and they move their limbs in the horizontal plane with the elbow and knee joints as fixed points. The body is therefore swung forwards by flexion of the shoulder and pelvic regions, and the neck, body and tail assist movement by their snakelike undulations. Most lizards are predatory, and relatively few feed on plants, but these are not pure vegetarians. Many lizards do not require drinking water but make do by licking dew from flat stones. When seized by the tail many species can break it off (autotomy), leaving the tail still wriggling, while they themselves escape. Some of the tail vertebrae have a cartilaginous disc inserted at their centre, and the muscles and blood vessels are so arranged that the break causes the least possible interference. Autotomy is discussed in greater detail under No. 60. A new tail grows out from the point of fracture, with a strand of cartilage replacing the vertebrae. The new tail seldom attains the length of the original. Lizards occur in all parts of the world except in the polar regions. They are commonest in the tropics, and decrease in numbers with distance from the equator.

46 Turkish gecko
Hemidactylys turcicus

Identification: 9–10 cm ($3\frac{1}{2}$–4 in.). This and the following three species belong to the gecko family, which includes the smallest of all reptiles. In most of them the lower eyelid has grown over the eye as a transparent capsule which is fused to the upper lid. The tongue is very long and it can keep the eyes clean and damp. Geckos are the only reptiles which have a true voice, and they can produce loud sounds. Most of them have an arrangement of small, transverse skin lamellae on the undersides of the toes which act as adhesive discs when pressed against the substrate. In this way they can run up vertical walls and along the ceiling. They nearly all lay 2 spherical, white eggs with tough leathery shells. The Turkish gecko has toes with short, pointed claws. The body wall is so thin that the eggs can be seen through it. The back is whitish, greyish or brownish-black, marked with irregular dark spots. The underside is uniformly white and covered with flat, hexagonal scales and the tail also has dark but otherwise similar scales running transversely.

Distribution: Along the coasts of the Mediterranean and on the islands. Also in northern Africa and western Asia.

Habitat and habits: Lives under stones, in rock crevices and in ruins, and is also very commonly seen in houses and cellars where at night it catches insects and spiders. The voice is much louder than that of the other European geckos.

47 Moorish gecko
Tarentola mauritanica

Identification: 12–18 cm ($4\frac{1}{2}$–7 in.). Head large and broad. Many of the scales on the upperside of the tail are strongly keeled, forming transverse and longitudinal rows of spines. Toes broad, only the third and fourth with claws. Back ash-grey to grey-brown with dark transverse bands which are particularly conspicuous on the tail. The body is usually darker in the sunshine and paler in the shade and at night. It is capable of changing colour relatively quickly.

Distribution: Countries of the western Mediterranean, eastwards to Dalmatia, the Ionian Islands and Crete, and probably commonest in Spain. Also found in the Canary Islands and Egypt.

Habitat: Mainly on rocks, walls and tree stumps, usually near the coast.

Habits: In contrast to most other geckos, which are essentially crepuscular or nocturnal reptiles, this species may be active both by day and by night. It has a very weak voice. The diet consists mainly of spiders and various insects.

48 Naked-fingered gecko
Cyrtodactylus kotschyi

Identification: 9–10 cm ($3\frac{1}{2}$–4 in.). Head large and flat, body very slender. Tail flattened and longer than the body. Fingers and toes only slightly flattened at the base where they have suction discs; at their tips they are laterally compressed and bent upwards. The colour of the back varies from pale grey to dark grey-brown with dark transverse bands, which are often bordered behind with white. Uniform grey specimens also occur. This species has a great capacity for changing colour and in sunshine it can, for example, become completely black.

Distribution: Southern Italy, the Balkans northwards to central Bulgaria and in Crete. Also found in western Asia and the Crimea. This is a very variable species and no fewer than 14 different subspecies have been described for Europe.

Habitat and habits: Lives under boulders and in rock crevices but if disturbed will usually run off and hide in the nearest thicket. Is most active during darkness, but can also be seen in broad daylight, although it always remains hidden during the hottest part of the day. When hiding under stones it has the habit of lying on its back.

Also known as *Gymnodactylus kotschyi.*

49 European gecko
Phyllodactylus europaeus

Identification: 6–7 cm (2¼–2¾ in.). Head large, body flat, and tail narrow at the root and thickest in the middle. The fingers and toes have flat tips with suction discs. Back greyish-yellow with dark transverse bands and black spots.

Distribution: Corsica and Sardinia and in a few places in southernmost France and Italy.

Habitat and habits: Essentially an animal of the twilight which does not tolerate strong sunshine. Found only out-of-doors, where it spends most of the time under stones or behind the bark of trees.

50 Mediterranean chameleon
Chamaeleo chamaeleon

Identification: 25–30 cm (10–12 in.). Body covered with small horny scales. Back raised and compressed to form a sharp ridge. At the back of the head there is a triangular helmet-like crest, which is largest in the male. Head large and broad with a long gape and a fairly pointed snout. Neck only marked by a constriction behind the head. Limbs slender and longer than the body. On the fore-limbs the 3 inner and the 2 outer toes are fused and they have transparent claws. On the hind-limbs the 2 inner and the 3 outer toes are fused. The feet are therefore somewhat like small pincers. The slightly compressed tail, which is a little longer than the body, can be coiled round a branch and is strong enough to support the weight of the body. The protruding eyes are covered with a thick eyelid, with only a small

opening for the pupil. The eyes can be moved independently of one another, so that a chameleon can look forwards with one eye and backwards with the other at the same time. The club-shaped tongue can be protruded and may then reach a length about the same as that of the body. The thin skin of the throat can be inflated.

Distribution and habitat: Found in southern Spain, and has been introduced into south Portugal, Crete and the Canary Islands. Also found in northern Africa, Asia Minor, Syria, Arabia and Cyprus. Although mainly an arboreal species it can also live in low bushes or on the ground in arid regions, where it may bury itself.

Habits: Chameleons move more slowly than any other reptiles, including the tortoises. They are best known for their ability to change colour, even though other animals, such as squid and cuttlefish, can do this far more efficiently. The colour can change from green to yellow, white or black and may be different on the right and left sides. The colour change is primarily dependent upon the lighting, although temperature and relative humidity are also important; there is also no doubt that psychical conditions play a part, for chameleons can, for example, turn pale when enraged. On the other hand, colour change does not play such an important role in camouflage against the background as was once thought.

After strenuous fighting among the males, chameleons mate in August–September and the females lay 25–40 eggs in a hole in the ground during September–October. The eggs do not hatch until the end of the following July. Diet consists of locusts, grass-hoppers, flies and many other insects. When the prey is sufficiently close, the chameleon slowly opens its mouth and very rapidly extrudes the tongue, whose sticky tip picks up the insect.

51 Slowworm
Anguis fragilis

Identification: 30–50 cm (12–20 in.); one of the longest recorded was 52·4 cm (21 in.). The long, cylindrical, snakelike body has no limbs and is the same thickness throughout; it passes without any constriction into a small head with rounded snout. The tongue, which is flat with a forked tip, can be protruded quite a distance and it also picks up scent particles and carries them back to Jacobson's organ, a sense organ in the roof of the mouth. The small eyes are furnished with eyelids. The ear cavities are usually not visible externally. The jaws have small, wedge-shaped teeth which are turned slightly backwards. The rear end of the body passes without any constriction into the cylindrical tail which is longer than the head and body combined. The tail ends in a short spine. The proportions are reversed in the newly born young, in which the tail is a little shorter than the head and body. On the top of the head there are several relatively large regular plates, but the remainder of the body is covered with small smooth scales. On the upper- and underside the scales are hexagonal and larger than on the sides where they are more rhomboid. In the newly born young the back is greenish, silvery or golden-brown, the rest of the body being bluish-black; a black spot on the nape extends back in the form of a narrow black stripe right along the mid-line of the back to the tip of the tail. With increasing age they become browner, until in the adults the ground colour may be like shiny polished copper. The black dorsal stripe becomes reduced with age, and the underside, which becomes paler, may often be completely steel-grey, a colour which can sometimes also be seen on the upperside. The majority of brown slowworms are females, most grey ones males. The males also have larger heads than the females. Specimens can be found with pale blue spots on the back, the incidence of these increasing towards the south-eastern end of the distribution range; this applies particularly to the males. Curiously enough the small external ear cavities are usually visible in the blue-spotted specimens. Slowworms have small bony plates in the skin which are closely attached to the scales, forming a flexible casing of dermal armour. There is fairly free movement between these plates. Slowworms move with serpentine undulation of the body, the underside taking advantage of every slight roughness on the surface of the ground. On a very smooth surface they will often hitch themselves along with the chin. When starting to move off they sometimes bend the tip of the tail like a hook and use the tail spine to gain purchase and then push the whole body forwards. In a similar way they can creep backwards by stretching out the tail, finding a firm point for the tail spine and then pulling the body backwards.

Distribution: This is the most widely distributed of all European reptiles. It occurs as far north as 64° N and is only absent from Sardinia and other islands in the Mediterranean, and from the Crimea and the Ukraine. Outside Europe it is also known from north-west Africa.

Habitat: Essentially a terrestrial animal which lives in damp, shady places, under stones or in burrows which it digs with the snout. Also found in deciduous woodland, gardens and fields. It is not uncommonly seen in ant-hills, where it is the only vertebrate present, and is also seen on sunlit walls and on woodland paths where it lies out to bask. In mountain regions it extends to altitudes of 2,000 m (6,540 ft).

Habits: Although very common, it is not often seen, for it lives hidden away and usually only comes out at twilight. It is also active after rain and possibly this is when it can find plenty of earthworms, which form an important part of the diet. Slowworms hibernate in the ground from October, and there may often be 20–30 gathered in the

same place, the entrance being blocked with earth and grass to keep out the frost. The warmth of the sun entices them out at the beginning of April and mating takes place in late April, May and June. The male grips the female behind the head with his jaws and twists his body in under her so that mating can take place.

Mating has seldom been observed probably because it most frequently occurs in secluded places, often underground. Very rarely one can see a male with the short, double copulatory organ extruded (this can be also sometimes be seen in specimens that have been killed), a phenomenon which has led to the erroneous belief that Slowworms have hind-legs.

Slowworms first become sexually mature at an age of 3–5 years. The young are born, usually below ground, in the latter half of August or at the beginning of September. The number of young is dependent upon the age of the female and may vary between 5 and 25, but is usually about 10. They are each born in a transparent, yellow egg membrane, which they quickly burst by rapid stabbing movements of the head, so it is correct to say that the Slowworm is ovoviviparous. At birth the young are 6·5–9·0 cm (2½–3½ in.) long in their northern range, but somewhat smaller to the south. They are very active as soon as they have left the egg membrane and it is not long before they begin to eat small slugs and thin earthworms. They first feel the prey with the snout, lick it with the tongue and then start to bite it, before swallowing it, a process which may take several minutes. When they become aware of a moving earthworm they will glide up and touch it with the snout, moving the tongue in and out.

They then lift up the front part of their body, bend the head downwards towards the worm and strike it with a quick bite. The earthworm twists itself round the Slowworm's snout which is kept pressed against the ground for some time, so that the head roughly forms a right angle to the body. It then usually retreats with the prey. As a rule, a Slowworm does not try to eat a worm that is more than half the length and thickness of its own body. If two Slowworms happen to seize a worm, one at each end, there may be a tug-of-war, until finally one of them wins and swallows the whole prey or else the worm is pulled apart. The slime from earthworms and slugs will often remain on the Slowworm's mouth, and it is cleaned off by gliding through the undergrowth.

Slowworms cast their skin about every six weeks, and the slough usually comes off in two or three pieces, only exceptionally in one piece. They can lose the tail even more easily than other lizards. They do not then grow a new tail but will grow a small, black, wedge-shaped stump at the point of fracture. There is no reliable information on the longevity of Slowworms in the wild, but there is a record of a specimen being kept in a vivarium for 54 years.

Slowworms are preyed upon by carnivores, hedgehogs and birds of prey, and occasionally Common toads will eat them.

52 Grey burrowing lizard
Blanus cinereus

Identification: 20–30 cm (8–12 in.). Head with a short, rounded snout and separated from the long limbless body by a transverse furrow. Tail short and cylindrical. Skin leathery and subdivided by shallow furrows into small rectangular areas. Eyes turned upwards and covered with skin. Colour of back varies from yellowish or reddish to brown. Underside paler.

Distribution and habitat: In Europe known only from the Iberian Peninsula, but also occurs in Morocco and Algeria. Lives in the ground under large boulders and quickly disappears in burrows which it has dug itself as soon as one lifts the stones.

Habits: In contrast to all other limbless reptiles the burrowing lizards move by means of vertical undulations of the body, which must be regarded as an adaptation to their underground, burrowing habits. They feed mainly on millipedes, ants, termites and other insects but also eat earthworms.

53 Fitzinger's lizard
Algyroides fitzingeri

Identification: 10–12 cm (4–4½ in.). Body slender, head flat and tail accounting for over two-thirds of the total length. Dorsal scales large and overlapping, with prominent keels. Corsican specimens tend to be larger than those from Sardinia.

Upperside olive-brown, throat pearl-grey, tip of snout blue, fore-legs greenish and the underside and hind-legs yellow.

Distribution and habitat: Found only in Corsica and Sardinia, where it is very common in certain places. Lives in among rocks and on walls, as well as under stones and tree roots.

Habits: These lizards are only seldom seen as they are shy and very fast. The female lays only 2 very large eggs.

54 Spinefoot lizard
Acanthodactylus erythrurus

Identification: 18–20 cm (7–8 in.). The popular name refers to the small fringed scales on the side of the toes. These fringes increase the area of the foot, thus facilitating movement across sand and they also render the feet better adapted for digging in sand. In this species the fringes are not so well developed as in related species which occur outside Europe. In young specimens the back is velvety-black with 7–9 yellowish longitudinal stripes. With increasing age these stripes become more conspicuous and small pale yellow spots appear. Old specimens have very large yellow spots on the flanks, and in the male there may also be blue markings. The young and the adult females have a characteristic red colour on the underside of the tail.

Distribution and habitat: Found in dry, stony and sandy regions in

Portugal, Spain and southern France; also in north-west Africa.

Habits: Very shy, normally hiding in cavities which it has itself dug. Usually only emerges to bask in the sun and to feed. The diet consists of grasshoppers and other insects.

55 Algerian sand racer
Psammodromus algirus

Identification: 20–27 cm (8–10½ in.). The large, pointed scales on the body are imbricated (overlapping like the tiles on a roof) and each has a sharp keel. The undersides of the fingers and toes are smooth, and the long thin tail accounts for over two-thirds of the total length. Upperside brown or olive-coloured, often with golden or coppery iridescence. Along each flank there are 2 yellow stripes with dark edges, and the back often has a narrow dark stripe. Underside white with a greenish tinge. The females have one or two small blue shoulder patches, and in the male these spots are larger, and there may even be several of them.

Distribution: Found in the Iberian Peninsula and on the Mediterranean coast of France. Also in northern Africa.

Habitat: Prefers dry, sandy or stony regions, but also climbs among low dry scrub or burrows into the warm sand when disturbed.

Habits: Difficult to observe as it is lightning fast and very shy. If caught, it squeaks, bites and slashes with the long tail. It also squeaks when agitated, particularly during fights in the breeding season.

56 Spanish sand racer
Psammodromus hispanicus

Identification: 10–12 cm (4–4½ in.). As in the preceding species the large, pointed and imbricated scales on the body have a sharp keel. The fingers and toes have small pointed scales on their undersides. The long thin tail accounts for less than two-thirds of the total length. The colour of the upperside varies from dark copper-brown to yellow-brown or grey. Young specimens usually have 6 whitish-yellow longitudinal stripes, those on the back gradually changing to pale spots with black edges. These spots become reduced in the adults, but they retain a whitish stripe on the sides, but sometimes these may also disappear, leaving the animal completely copper-brown or grey. The underside is shiny pearl-grey, often shading into brownish or greenish. During the breeding season the males have two pairs of

conspicuous blue shoulder spots with white edges and also a row of small blue spots along the sides, bordering the belly.

Distribution: Two races, one in the west and south of the Iberian Peninsula, the other in eastern Spain, extending into the coastal region of southern France.

Habitat and habits: Lives in similar places to the preceding species. Only really active when the sun is shining and enters hibernation early, burrowing into the sand. In June the female usually lays 6 white eggs deep down in the sand and the young hatch out at the end of July or beginning of August. During the spring one only finds small or half-grown specimens, which suggests that they only live for 1 year and die during their second winter.

57 Round-bodied skink
Chalcides bedriagai

Identification: 11–12 cm ($4\frac{1}{4}$–$4\frac{1}{2}$ in.). Body long and cylindrical with very short limbs. The fore-limbs are particuiarly short and they can be withdrawn into grooves on the sides of the body. The tail, which is about the same length as the body, is broad and flat at the root, tapering to a point. The scales are nearly all the same size and completely smooth. As in all the other species in the skink family, the central part of the lower eyelid is transparent, so that the animal can see even when the eyelids are closed, thus preventing sand from entering the eyes. The upperside of the body appears to be uniform pale or dark bronze, the sides somewhat darker, but in fact both areas have a scattering of small black

oblong spots. These are larger and closer together on the sides where they sometimes form what are almost longitudinal stripes. The underside is pale grey and the tail is yellowish underneath.

Distribution and habitat: Found in the mountain regions of the Iberian Peninsula where it lives under rocks or buried in the ground.

Habits: Only emerges when the sun is shining, or when searching for food, but otherwise very little is known about its habits.

58 Sand skink
Chalcides chalcides

Identification: 24–40 cm ($9\frac{1}{2}$–15 in.), exceptionally somewhat longer. Body long, cylindrical and snakelike, the same thickness throughout, passing without constriction into a relatively small head with rounded snout and at the other end into a long tail which tapers gradually to a point and which accounts for half the total length. The very small, rudimentary limbs have only 3 toes and the tiny claws are almost invisible. The hind edges of the smooth, regular scales are arched but not keeled. The ground colour of the upperside is most frequently shiny silvery-grey to olive-green, but may also be brown. The back normally has 2 pale stripes with dark edges and there are 2 dark stripes along each side. One of the races may have 9–11 longitudinal stripes, but unstriped specimens also occur. The colour of the underside may be bluish, grey, pale brown or olive-green.

Distribution and habitat: The Iberian Peninsula, southern France, Sardinia,

Sicily, Elba and Italy, and also occurs in north-west Africa. Lives mainly in damp areas in the vicinity of the coast.

Habits: Like the preceding species this skink produces live young. It can move fast through the grass when hunting insects, spiders, worms and small slugs. It burrows into the ground during the cold part of the year.

59 Sand lizard
Lacerta agilis

Identification: 15–20 cm (6–8 in.), exceptionally up to 24 cm (10 in.). Body slender in the young but becoming stouter with age. Snout short and the tympanic membrane, looking like a dark brown patch, lies just behind the eye. Mouth broad, with small pointed teeth on the jaws. Tongue long, narrow and forked at the tip. Claws on the fore-limbs at least half as long again as those on the hind-limbs. Tail elongated, thick at the root, flat on the upperside and one and a half times the length of the body. Scales on the head relatively large and flat, lying close to each other. Along the middle of the back the scales are narrow and overlapping, with keels. On the sides the scales become broader and flatter. On the legs the scales are round and granular, and on the tail overlapping and arranged in rings. Coloration and pattern is dependent upon age, sex, season and locality. The young are usually grey-brown or yellow-brown with a varying number of white eye-spots with black borders arranged in irregular longitudinal rows. With increasing age the two bands on the back become paler and the eye-spots fewer and larger, usually leaving one row along the dorsal line and one or

two along each side of the body. The females have the most conspicuous pattern but the males have the most handsome colours; in the mating period they show a beautiful green along the sides of the body and usually also on the head and legs. Underside pale grey in the juveniles, yellowish-green or pale blue with black spots in the adult males and whitish or sulphur-yellow usually without spots in the females. The male has stouter hind-legs and a thicker tail root than the female.

The genus *Lacerta* is by far the largest reptile genus in Europe, containing some 30 species, many of which have several subspecies. They are characterized by having well-developed limbs, a long pointed tail, visible tympanic membranes and a cleft tongue. The belly has small, square horny scales and the scales on the tail are arranged in transverse rings. They feed only on live food.

Distribution: Widespread over the greater part of northern and central limit, with the northern limit running from southern England through the northern part of Denmark to southern Sweden and farther eastwards. To the south this species reaches the Alps, Hungary, Czechoslovakia, the northern Balkans and eastwards to the Black Sea.

Habitat: Lives on stone walls, heathland and other sun-baked places, where the soil is loose and without too much vegetation; also common on sand dunes. It can dig holes, but often lives in old mole runs and field mouse burrows, to which it retreats in bad weather and at night. The entrances to these shelters are often blocked with grass, moss or earth as a protection against the cold.

Habits: Enters hibernation in September–October and emerges again in April, the males often 2–3 weeks before the females. During the warmest part of the year Sand lizards may retire to cool hiding-places and aestivate. Like their relatives in the genus *Lacerta* they orientate themselves by flicking the tongue in and out, picking up scent particles which are transferred by the cleft tip to Jacobson's organ which is situated in the roof of the mouth. This organ's olfactory epithelium plays a part in tracking down prey and in initiating mating. Before mating starts in spring the males fight violently over the females, and one may even bite off its opponent's tail. But it is not only during these mating season fights that they show their aggressiveness. They often bite when caught and can seize a finger so firmly that one can lift them off the ground. Mating usually takes

place at the beginning of May and the 'gestation' period lasts about 6 weeks. Towards the end of this period the female becomes so stout that one can see the outline of the eggs inside the body. She digs a hole in the ground using her fore-legs, and lays in it 5–13 oblong white eggs with soft shells. The eggs increase in size after they have been laid, probably because they take up water. Development starts before the eggs are laid, and when laid the embryos are about 1 cm ($\frac{1}{3}$ in.) long. The eggs hatch in 6–8 weeks, the young lizard tearing a hole in the tough eggshell with its sharp egg-tooth, a temporary structure situated on the tip of the snout. Details of sloughing, tail autotomy, diet and enemies are given under the next species.

60 Viviparous lizard
Lacerta vivipara

Identification: 10–16 cm (4–6$\frac{1}{4}$ in.), exceptionally up to 18 cm (7 in.). Body slender, particularly in the male, head relatively long with a rounded snout. The large tail is almost the same thickness along its first half and then tapers to a point. In the female the tail is about the same length as the body, but in the male it is somewhat longer, and is thicker at the root. Legs relatively long, with the dark claws the same length on the fore- and hind-limbs. The horny shields on the head are relatively large, and the scales on the back are elongated, hexagonal, overlapping and with prominent longitudinal keels. On the underside the shields are square and arranged in 6–8 rows, those in the central rows being the smallest. On the tail the uppermost scales are pointed with a very promi-

newly hatched young are always very dark, often nearly black, but usually have 2 rows of pale spots down the back.

Distribution: This is the most widespread of all the European species of *Lacerta* and its range extends farther north than any of the others—in Scandinavia right up to 70° N. It occurs over practically the whole of northern and central Europe, and to the south it extends to the northernmost parts of the Iberian Peninsula, to Italy and the Balkans. Also found in northern Asia eastwards to the Amur and northern Mongolia.

Habitat: Lives mainly in damp places, in meadows and moorland and on the outskirts of woodland. Is common in most localities but does not occur in very large numbers as do certain other species of *Lacerta*. Often sits and basks on a tree stump or a rock, especially in the morning sun and in the afternoon. In bad weather and at night it retreats beneath the bark of trees or under moss and other protective plants. It can tolerate cold but not tropical or subtropical climates. In the northern part of its range the Viviparous lizard lives in low-lying country, often on sand dunes but farther south can be found at altitudes up to 3,000 m (9,800 ft). A very stationary animal which does not stray far from its hiding-place, to which it will retreat when disturbed. It does not run particularly fast nor very far at a time, so it is not very difficult to catch and it does not usually bite. However, it is more difficult to catch when close to water, as it jumps in and swims remarkably well, bending the body and tail and holding the limbs against

nent keel, on the underside of the tail root they are smooth with a rounded posterior edge, and towards the tail root they are all pointed and keeled. Neither the colours nor patterns vary so much as in most other European species of *Lacerta*. The upperside varies between grey-brown and red-brown, broken up by pale or dark spots which are arranged in more or less distinct longitudinal rows. In most specimens there is a row of black spots running along the mid-line of the back which may join up to form one or more longitudinal stripes. Along the sides there are often longitudinal rows of white or yellow spots, sometimes with black spots as well. In the males the underside is usually reddish-yellow with numerous black dots, whereas in the females it is bluish-grey or yellowish-white, without dots. Exceptionally, completely black specimens (melanistic varieties) occur in this and also in other species of *Lacerta*. The

the flanks. It can also dive and hide in the mud at the bottom.

Habits: Enters hibernation during October, going into holes in the ground or under tree stumps. Emerges again at the end of March, which is earlier than in other species of *Lacerta.* During the period of hibernation it loses 2–10 per cent of its body weight. In very warm weather it may aestivate. Mating takes place in May. The pair mates for up to 30 minutes, then the female starts to move away, after releasing herself from the male who is holding on to her with his teeth. It is not until the female starts to twist and turn that the male eventually lets go, and the two lizards slowly move away from each other. Blue horseshoe-shaped marks on the female's belly bear witness to the bites of the male during mating; these marks may remain visible for several months. The period of gestation is almost 3 months and the young are usually born at the end of July, for this species, unlike its relatives, produces live young. In fact, at birth the young are enclosed in a thin, whitish egg membrane. This is broken open by sideways movements of the head and the youngster then creeps out. Exceptionally, the membrane bursts in the oviduct, but more usually the young are encased in the membrane at birth; the former phenomenon has led to the view that this species lays eggs (or is oviparous). But an egg can be defined as an embryo enclosed in a shell which undergoes development within this shell, outside the body of the mother. It is, in fact, more correct to say that *Lacerta vivipara* is ovo-viviparous. The difference between ovoviviparity and true viviparity

is that in the latter case the developing embryos receive nourishment from the mother's body through a special structure which transports food from the blood of the mother to that of the embryo. In the so-called live-bearing or viviparous reptiles, such as the Slowworm and the Adder, the mother acts purely as an incubator and the embryos are nourished entirely by their own large yolk sacs, except that they obtain water from the mother. Before the birth the Viviparous lizard female seeks shelter under withered plants or in similar places and then usually produces 6–8 young; the number does, however, vary from 5 to 12. The birth of each young one only lasts a few seconds, and the whitish egg membrane is so thin that the 'egg' appears blue-black owing to the dark colour of the young lizard inside. The egg membrane is on average 11 mm ($\frac{7}{16}$ in.) long and 8 mm ($\frac{5}{16}$ in.) broad. There is usually an interval of 5–8 minutes between each individual birth, so it takes about an hour for the whole litter to be produced. When finished the female has nothing more to do with the young, which are then on average 38 mm ($1\frac{1}{4}$ in.) long and can run about shortly after birth. The little yolk sac which is attached by a thin thread of tissue is quickly rubbed off as the young one begins to move around. In the Pyrenees this species evidently does lay eggs which are in varying stages of development. After giving birth the female becomes relatively slender.

Viviparous lizards moult or slough at intervals of 1–2 months during the period when they are not hibernating. Evidently the females do not slough so frequently as the males, possibly because they are not so active, particularly during the gestation period. The slough strips off, and sometimes forms a sheath from the neck to the root of the tail. Usually, however, the horny cuticle bursts along the back and is soon hanging in shreds, which the animal gets rid off by rubbing against stones and plants or by pulling it off with the mouth. When pursued these lizards can run very fast and if seized by the tail will break it off, leaving it to writhe while the lizard itself escapes. At one time it was thought that the tail was broken off and left to writhe on the ground in order to distract the enemy for a few valuable seconds but there is really no evidence for this. Nor is it necessarily the result of too heavy a tug because, one can, for example, lift a tame lizard off the ground by the tail without it breaking. The breakage process, known as autotomy, is due to a sudden and powerful muscular contraction which is independent of whether or not the lizard is in the hands of an enemy. Each of the much elongated tail vertebrae has a transverse breakage plane running across it, where ossification is incomplete, and it is at one of these places that the break occurs. The relevant muscles and blood vessels are so arranged that the break causes the minimum amount of interference. A new tail grows out from the breakage point. Autotomy does not take place at any fixed point along the tail. Exceptionally, it may occur near the root of the tail or very close to the tip, but normally it happens about half-way along. When the wound scab falls off 2–3 weeks later an underlying layer of skin has formed which is dark above and pale below. Soon transverse furrows and later longitudinal folds

appear on the new tail and in the course of about 3 months it is not very different in colour from the old one. The new tail does not attain the full length of the old one, is not so scaly and it sloughs independently of the rest of the animal. In this reconstituted tail the vertebrae are replaced by a strand of cartilage, so this is not really a case of true regeneration, but rather of a repair. Abnormal cell divisions may cause the production of two tails instead of one or the old tail may only break off half-way across, when a new tail will grow out from the damaged place. This ability to reconstitute tissue is only associated with the tail. A toe bitten off never grows again, still less a limb.

Viviparous lizards feed mainly on earthworms, insects and larvae, but they will in fact eat anything living of a suitable size and occasionally even young of their own species. Large prey is shaken and banged against the ground, then placed long ways on so that the lizard can swallow it. Irrespective of its size the prey is never bitten, but always swallowed whole. The Viviparous lizard has many enemies, including foxes, martens, crows and many birds of prey.

Also known as the Common lizard.

61 Wall lizard
Lacerta muralis

Identification: 18–20 cm (7–8 in.), very exceptionally up to 25 cm (10 in.). Body slender, head very long and pointed, and a tail that accounts for about two-thirds of the total length. Back usually brown or grey but it may sometimes be almost completely black. Adult males may have black spots or a dark, reticulated pattern on

the upperside, females a row of dark spots along the mid-line, and along each side a similar row of spots which are edged with white above and below. Underside whitish, yellowish or reddish, sometimes with black spots. There may be a row of blue or black spots along the edge of the belly. The coloration is so variable that it is often difficult or impossible to say whether a specimen is a Wall lizard or whether it belongs to one of the following six forms. Even experts may be uncertain or disagree, and in many cases it is only the geographical distribution which determines the identification. The colour plates therefore only show a typical male and female Wall lizard and none of the closely related forms.

Distribution: Very common in southern Europe, extending northwards to France, Belgium, the Netherlands and western Germany. Over the whole range it occurs rather irregularly,

often in isolated populations. Also found in Asia Minor.

Habitat: Found mainly in dry, sunny, rocky areas and often seen on cliffs and garden walls, sometimes being quite abundant inside small towns, and it may also occur in dry, open woodland.

Habits: Very fast and can climb up vertical walls. Very hardy and the period of hibernation is short. In May, June or July the female digs a hole in the ground and lays several clutches of 2–8 eggs.

Lacerta hispanica

Identification: 14–15 cm (5½–6 in.). Body slender and flat, snout pointed. Back grey, brownish or olive-coloured, sometimes faintly greenish. The male has black longitudinal stripes along the back which are often mixed with small pale spots. In the females and juveniles the black longitudinal stripes have white edges. Tail with transverse blue and black bands.

Distribution: Spain, in the mountain regions along the Mediterranean coast between Alicante and Malaga. Also found in north-west Africa.

Lacerta monticola

Identification: 18–19 cm (7–7½ in.). Very similar to the Wall lizard but with a flatter body and a more pointed snout. Upperside usually olive-brown, sometimes reddish-brown, with a double row of irregular markings, which continues out on to the tail as 3 rows of dark brown spots.

Distribution: In the mountains in the north-western part of Spain and in north-western Portugal, extending up to altitudes of 1,500–1,800 m (4,900–5,870 ft).

Lacerta bedriagae

Identification: 15–20 cm (6–8 in.). Front of head very short but also very pointed. Upperside greenish with numerous black spots or irregular narrow transverse bands. Sometimes the green colour is almost completely covered with dark spots and bands. There are often blue spots on the sides of the body.

Distribution: In the mountains of Corsica and Sardinia at altitudes of 700–2,700 m (2,300–8,940 ft), living mainly in forests and in the vicinity of water.

Lacerta tiliguerta

Identification: 16–18 cm (6¼–7 in.). Very similar to the Wall lizard, but with a broader and shorter head. Coloration varies considerably, the upperside from pale green to brownish, the underside from pale grey to yellowish or reddish. The tail is green. There are 4 pale longitudinal stripes on the upperside and pale blue shoulder markings.

Distribution: Corsica and Sardinia.

Lacerta wagleriana

Identification: 16–25 cm (6¼–10 in.). Back almost uniformly green with a few dark spots arranged in longitudinal rows. Sides of the body brownish with darker spots.

Distribution: Found in the western, central and southern parts of Sicily and on the Lipari Islands.

Lacerta filfolensis

Identification: 16–25 cm (6¼–10 in.). Very similar to the Wall lizard, but larger and with a short, pointed snout. Tail two-thirds of the total length. Body completely black, except for the underside of the tail which is grey.

Distribution: Malta and the neighbouring islands, including Filfola.

62 Ruin lizard
Lacerta sicula

Identification: 20–30 cm (8–12 in.). A relatively large body, a broad pointed head and a long tail which accounts for about two-thirds of the total length. Like the preceding 5 species and the Wall lizard the coloration is very variable, and in fact about 40 races or subspecies of this lizard have been recognized, more than in any other reptile species in Europe. In the Italian form, which is by far the most abundant, the upperside is usually greyish-yellow or blue-green with black spots arranged in longitudinal rows. On the sides the spots are

irregular and they start at the shoulder with a sky-blue marking. The mid-dorsal stripe, which varies considerably in breadth, runs right along the back and out to the tip of the tail. The underside of the body is whitish, yellowish or faintly greenish, and in the males it is sometimes brick-red. The males also have blue spots marked with black dots along each side of the body. Some specimens have almost no pattern and others are completely black. There is a blue form living on two small cliffs at Capri.

Distribution: Most numerous in southern Italy. Widespread throughout Italy, Corsica, Sardinia, Sicily, the western Balkans and along the shores of the Sea of Marmora.

Habitat: Lives at the foot of cliffs, on walls, on flat roofs of houses and among the ruins of ancient buildings. Found, for instance, among lava blocks on Mount Vesuvius and in churchyards.

Habits: Just as fast as the Wall lizard and as difficult to catch.

63 Green lizard
Lacerta viridis

Identification: 30–40 cm (12–15 in.). Tail about a half to two-thirds of the total length. Upperside of male shiny yellow-green or grass-green, often with numerous black spots. Underside pale yellow and in the breeding season the male has a blue throat. He also has a larger head and a thicker root to the tail than the female. On the other hand, the females usually have a brown pattern on the upperside and 2–4 whitish or yellowish more or less continuous longitudinal stripes. Juven-

iles are brownish on the back and green on the sides, sometimes with pale stripes. Coloration varies not only with age and sex but also with the locality.

Distribution: Throughout the whole of central Europe from northern Spain in the west to southern Russia in the east, northwards to Paris, the Rhine, south Germany and Poland. In the southern part of the range there are various races in Italy and the Balkans. Also occurs in Asia Minor and Iran.

Habitat: Lives mainly in dry areas with scrub and bushes or on heathland, and sometimes on the outskirts of coniferous forests. Found principally in low-lying country but in mountainous regions up to altitudes of 1,500 m (4,900 ft). Normally, Green lizards are very stationary but from time to time some of them may move about, perhaps over fairly long distances.

Habits: Climbs fast and surely among rocks and bushes and if threatened will jump into the water or climb high up into a tree. If they cannot escape they will open the mouth wide and prepare to bite. During the mating season the males fight strenuously over the females. Six to eight weeks after mating the female lays 5–20 yellowish-white eggs which are covered up with a small heap of damp sand. The eggs hatch in about a month and the young are 8–9 cm (3–3½ in.) long.

Green lizards feed mainly on insects and larvae but they will also eat other lizards and their eggs, as well as slowworms and small snakes. In the northern part of the range they go into hibernation from October to April, but in the south they are active throughout the year.

64 Schreiber's lizard
Lacerta schreiberi

Identification: 25–30 cm (10–12 in.). Very similar to a medium-sized specimen of the preceding species. Head very large. The long thin tail accounts for about two-thirds of the total length. Upperside pale olive or grass-green with large black spots which tend to form one or two longitudinal rows along the mid-line of the back. There are also black patches on the sides; these become smaller towards the belly which is nearly always spotted with black. The tail has dark transverse bands which are larger in the female than in the male. In juveniles the back is uniformly brown and each side of the body has 2–4 rows of spots with dark edges, a colour pattern which is sometimes retained in the adults. Specimens in which the large dark spots on the upperside are replaced by numerous small black dots are very similar to the Green lizard, but they can be distinguished by their black-spotted undersides.

Distribution and habitat: Found only in the western parts of the Iberian Peninsula. They live in among boulders, on tiled walls and in woodland and may occur at altitudes up to 800 m (2,600 ft).

Habits: Like the preceding species this is a fast climber which moves about with confidence on rocks and among bushy vegetation. It is not uncommon for the pairs to live isolated during the mating period, but usually the males indulge in strenuous fighting. The female lays 5–20 eggs which hatch in about a month. Diet consists mainly of insects and larvae but like its larger relative, the Green lizard, this species also feeds on other reptiles, such as lizards, Slowworms and small snakes.

65 Eyed lizard
Lacerta lepida

Identification: 50–60 cm (20–23 in.), exceptionally even longer. Head and body strikingly large, particularly in old individuals, and the tail is two-thirds of the total length. Head brownish above and green on the sides. Back brownish with a network of green and yellow lines. Sides green with 3 or 4 longitudinal rows of large blue markings with black borders. Underside pale yellowish-green or yellowish-white and without markings. In very large specimens the head and the front part of the body are uniform in colour without pattern. Juveniles grey, later acquiring large olive-brown patches which have whitish or bluish-black borders.

Distribution and habitat: Iberian Peninsula, southern France and Liguria in Italy. Also occurs in north-west Africa. Usually lives down on the ground, but sometimes climbs into bushes and trees. In the warm summer months it goes down to dried-out river beds.

Habits: Runs and jumps with great ability, and indeed it can make horizontal jumps of up to 1·5 m (5 ft). Like the 2 preceding species it can run up into trees and jump down from the top, even from heights of 5 m (16 ft), when wishing to escape. However, when faced with a dog or a cat it will defend itself and bite them in the muzzle or throat. They often live in pairs, even after the mating season, and the Green lizard also does this. The males fight savagely over the females. Mating normally takes place in May and 6 weeks later the female lays 6–10 eggs which are hidden in hollows trees or in holes in the ground. The eggs hatch in about 90 days.

Eyed lizards feed on insects, mice, small birds and snakes and other lizards. They also eat cherries, grapes, figs and other fruits.

Snakes

There are about 2,700 species of snake, of which only 33 occur in Europe. They have a very elongated body, a relatively short tail and no limbs. Snakes lack a sternum, urinary bladder, limb girdles, ear-drum and external ear opening. Most of the body is covered with horny scales; the underside has wide, transverse scales known as ventral plates. The jaws and facial bones are only loosely joined to each other and the two halves of the lower jaw are connected by a ligament and not fused, so that they can be moved independently of one another; this allows snakes to swallow relatively large animals. The tongue is forked and protractile.

There are slender teeth, pointing backwards, on the upper jaw and sometimes also on the palatal bone. The fangs of venomous snakes, which are only in the upper jaw have a frontal groove or a canal, through which the venom can flow. When a venomous snake bites or 'strikes' the venom glands are automatically emptied, the venom passing down the fangs to an opening just above the tip. Of the species described here the Montpellier snake and the 5 vipers are venomous.

Snakes use the tips of the ribs in locomotion, rather as though they were millipedes enclosed in a sheath of skin. As soon as a snake is born or hatched it immediately moults or 'sloughs' and this process is repeated several times a year throughout life. The horny cuticle becomes loosened along the jaws and the snake crawls out of the slough or rubs it off against stones and branches. The first sloughing of the year takes place when the snakes come out of hibernation in April–May and afterwards they slough about once every month until they go into hibernation in the autumn.

66 European whip snake
Coluber viridiflavus

Identification: 150–190 cm (60–75 in.). This very slender snake is easily recognized by its black upperside which has yellowish-green spots, forming narrow transverse bands on the front part of the body and longitudinal stripes on the rear part. The underside is almost greyish-yellow.

Distribution: Found in north-east Spain, southern France, southern Switzerland, north-western and southern Italy, Corsica, Sardinia, Malta and Elba.

Habitat: Prefers sunny areas with bushy vegetation, stone walls and vineyards.

Habitats: An aggressive snake which moves very fast. It also swims well and climbs with great agility among bushes to raid birds' nests. It feeds, however, mainly on lizards and also eats small rodents, frogs, grasshoppers and other insects, and even small snakes. The prey is killed by constriction if it is large, or just seized and swallowed if small.

67 Horseshoe snake
Coluber hippocrepis

Identification: 100–175 cm (40–68 in.). Very slender, with the tail accounting for about a fifth of the total length. The dark upperside of the head has a semicircular marking, which accounts for the name. The upperside is blackish, often with purplish or bluish iridescence, and with rows of small yellow or reddish markings. These spots are largest along the middle of the back, where they are almost oval in outline, but are smaller on the sides. The underside varies from whitish or yellowish to red.

Distribution: Widespread in the Iberian Peninsula and possibly also in Sardinia. Also occurs in north-west Africa.

Habitat and habits: A very active and aggressive snake which lives in holes in the ground or among rocks on dry hillsides. Also frequents houses where it hunts rats and mice in the cellars or climbs up the walls to raid sparrows' nests.

68 Leopard snake
Elaphe situla

Identification: 80–110 cm (30–40 in.). A relatively slender snake with a very short tail which is scarcely a sixth of the total length. The ground coloration is pale yellow or reddish-grey on the back, brown on the sides and greyish on the underside. The attractive circular leopard spots on the upperside have black borders and are sometimes fused in the middle line, so that they form a double row down the back. Along each side there is a row of smaller and darker spots which are positioned in the spaces between and below the large ones. Sometimes the large spots fuse to form long stripes which extend out to the tip of the tail. Towards the rear the ventral shields are marked with dark spots which increase in size towards the tail.

Distribution: Widespread in south-east Europe, particularly in southern Italy, Sicily, the Balkans, Crete, extending to the Caucasus, Crimea and Asia Minor.

Habitat and habits: Lives in sandy and stony areas with sparse bushy vegetation and old walls. Feeds almost exclusively on mice.

69 Ladder snake
Elaphe scalaris

Identification: 130–150 cm (52–60 in.). In this species the transition from head to tail is only faintly marked. Tail scarcely a seventh of the total length. In juveniles the ground colour of the upperside varies from pale grey to grey-brown, in adults from reddish-yellow to reddish-brown. The young have broad dark brown transverse bands (the rungs of the 'ladder') joined by 2 longitudinal stripes. There is usually 1, and sometimes 2 or 3 rows of small dark spots along the sides. This pattern disappears with age until only a trace of the cross bands remains, although the longitudinal stripes are still quite clear.

Distribution: Found in Portugal, Spain, Minorca and also along the Mediterranean coast of France.

Habitat: Prefers dry, sunny areas with hedges and vineyards, where it spends the day out in the open, and even on very hot days it can often be seen coiled up or actively crawling around.

It usually spends overcast days under rocks, in hollow trees or in holes in the ground.

Habits: A very fast and agile, diurnal snake which is an excellent climber. It feeds mainly on field mice, but also takes many small birds, lizards and grasshoppers. Mating takes place at the end of May or the beginning of June and 20–25 days later the female lays about 10 elongated, white eggs, which hatch in September–October. In contrast to the adults the newly hatched young move very slowly.

70 Four-lined snake
Elaphe quatuorlineata

Identification: 180–225 cm (70–90 in.). One of the longest of the European snakes. The Four-lined snake changes its external appearance to a remarkable degree during the course of its life. The young are pale grey with dark brown spots. They have a large dark spot on the back of the head and a dark stripe running from the eye to the corner of the mouth. The upperside has 5 rows of dark brown patches, the whitish underside has a number of triangular black spots. With increasing age the markings become paler, but the black stripe from eye to mouth remains black. There are 4 longitudinal stripes, which are eventually completely black. The ground colour also becomes an attractive dark brown.

Distribution: Found in central and south Italy, in Sicily and the western and southern parts of the Balkans. Also occurs in western Asia.

Habitat and habits: Frequents rocky areas as well as woodland and the banks of rivers. This is a good climber with a powerful body, but has a peaceful temperament and in captivity it seldom bites. The young feed mainly on lizards, but the adults also eat rats, mice and small birds.

71 Aesculapian snake
Elaphe longissima

Identification: 100–200 cm (40–80 in.). This species is named after Aesculapius the legendary Greek god of medicine, and in ancient times it was regarded as a sacred animal. The body is slender and there is no distinct constriction between the body and the narrow oblong head. The slender tail accounts for a fifth of the total length. The upperside is usually olive-brown, the head and neck often straw-coloured; many of the scales have thin white borders. In juveniles the sides of the head are yellowish, but this tends to disappear with age. In general, the

young snakes are darker than the adults. The underside is usually yellowish-white and may be spotted. But the coloration varies considerably and may be straw-coloured, grey, brown, blackish-green or almost completely black. The ventral plates have sharp edges which produce a ridge along each side of the belly. These ridges or keels enable the snake to get a grip on crevices and uneven places when climbing.

Distribution: The centre of the range is in Italy, but this snake also occurs in scattered localities in north-east Spain, France, Switzerland, Austria, Czechoslovakia, Poland, Hungary, and the Balkans. In Germany it is found in the vicinity of Schlangenbad and at Passau. The species also occurs in Asia Minor and northern Iran. To explain the scattered distribution in central Europe it has been suggested that the Romans travelled with these snakes, as sacred animals, which were later released. But it is more likely that these scattered occurrences represent the last relics of a former continuous distribution.

Habitat: Unlike most other snakes, the Aesculapian snake is not particularly tied to sun and warmth, and it can, for example, be seen on moonlit nights when out hunting for mice. Lives mainly in open deciduous woodland, in old walls and stony fields, and also in middens and stables where it sometimes lays its eggs and spends the winter.

Habits: Does not move particularly fast on the ground but is better than any other European snake at climbing, and has great muscular power. When coiled round a branch it is almost

impossible to unwind it. When climbing it takes the eggs and young from birds' nests, and in fact feeds mainly on small warm-blooded animals, but will also eat lizards. However, the main part of the diet consists of mice. It first seizes these with the fangs and then, quick as lightning, coils the front part of its body around the prey, which is then crushed. Soon afterwards, the snake releases its grip, takes hold of the mouse's head and swallows it, a process which takes a few minutes. It comes out of hibernation at the end of May and mates soon afterwards. The female lays usually not more than 5—8 eggs in July; these are not unlike giant ant pupae in shape and general appearance. The young hatch out in September and they are then about 12 cm ($4\frac{1}{2}$ in.) long. Shortly after this both young and adults go into hibernation.

72 Smooth snake
Coronella austriaca

Identification: 55—75 cm (22—29 in.), exceptionally 85 cm (32 in.), the female being larger than the male. A slender snake with smooth, shiny scales. The head is relatively small with a rounded snout and small eyes with round pupils and a yellow iris. The tail is relatively short and accounts for a fifth to a sixth of the total length. The generic name, *Coronella*, meaning a little crown, may refer to the large horny plates on top of the head or to the characteristic dark marking at the back of the head. The ground colour of the upperside is usually brown, but may be grey, yellowish or reddish and is often darker along the mid-line than on the sides. The top of the head is always dark brown, and a dark stripe runs from the nostrils through the eye to the corner of the mouth. The 'crown' marking on the back of the head is continued along the whole of the upperside as a double row of oblong, dark brown spots and the same applies to the eye-line, although this is not so distinct and continuous. Two or more spots in the same row may fuse to form stripes. In general, there is considerable individual variation, both in colour and pattern. In the juveniles the spots are nearly always darker and more distinct, and they may be completely black. The underside is uniformly reddish in the young, but brownish or greyish-yellow in the adults, sometimes with conspicuous dark spots. The area around the mouth and the underside of the head is pale brown.

Distribution: Mainly in central and south Europe, but also found as far north as Stavanger and Oslo in

southern Norway and in parts of south Sweden. To the south it is known from the northern half of the Iberian Peninsula, Italy and the Balkans. Also found in the Caucasus and Asia Minor, and in a restricted area in southern England (isolated localities in Dorset, Wiltshire, Hampshire, Surrey and Sussex).

Habitat: Essentially a diurnal animal, which usually lives in sunny places with a dense growth of heather and bushes. In the mountains it may occur at altitudes up to 2,000 m (6,540 ft). Very stationary in its habits, only leaving its shelter to hunt for prey.

Habits: Rather slow in movement and not a good climber, but swims well. When prevented from making its escape it may coil itself up and hiss, but there is great individual variation in what takes place in such a situation. It feeds almost exclusively on lizards, Slowworms and small snakes, but occasionally may also take mice and young birds. It seizes prey by the head, envelopes it in two or three coils and slowly swallows it, a process which may take half an hour. It only holds the prey firmly when coiled round it and does not suffocate it by constriction as some other snakes do. It usually comes out of its 6-month period of hibernation in April and mates immediately afterwards. The male seizes the female's tail in his mouth, holding on in the same way as lizards do. In August–September the females produce 3–15 young, each enclosed in an egg membrane from which it immediately breaks out; they are therefore ovoviviparous. The newly born young are 15–20 cm (6–8 in.) long. Smooth snakes go into hibernation at the end of September or the beginning of October.

73 Southern smooth snake
Coronella girondica

Identification: 60–70 cm (23–27 in.). Usually thinner and more slender than the Smooth snake, with a rather more elongated head. The tail which is also thinner accounts for a quarter of the total length. The ground colour of the upperside is pale or dark yellow-brown or grey-brown and may grade into reddish. The black stripe from the rear edge of the eye to the corner of the mouth usually continues down along the side of the neck. The marking on the nape is usually in the form of a horseshoe with the free ends turned backwards. The underside is sulphur-yellow or orange-red with 2 rows of square black markings or two longitudinal black stripes. The young are pale with 2 rows of dark spots

on the back and a bright red underside.

Distribution: Spain, Portugal, southern France, Sardinia, Sicily and Italy. Also found in north-west Africa.

Habitat and habits: Lives on dry slopes, walls and in fields in low-lying country and among hills. Shelters in deserted mole burrows, under stones and piles of timber and in winter in compost heaps and middens. This is a crepuscular snake which is seldom seen in full sunshine. It prefers to come out after sunset or on moonlit nights. Its feeding and breeding habits are the same as those of the Smooth snake.

74 Viperine snake
Natrix maura

Identification: 80–100 cm (30–40 in.), the male not more than 80 cm (30 in.). Body powerfully built, head broad with a short snout. Eyes facing slightly upwards. The relatively pointed tail accounts for a fifth of the total length. The ground colour and pattern of the upperside vary considerably according to age and locality, and may be grey-brown or olive-green with a more or less yellowish tinge. The upperside of the head has two square markings and along the back there is a row of more or less continuous dark patches, which often form a narrow zigzag stripe; this is why this snake is often confused with the Adder. The zigzag stripe is most frequent and most conspicuous in young specimens. Along each side of this stripe there is a row of large dark eyespots with white or yellow centres. The underside is grey-green, yellowish or reddish with pale or dark spots.

Body very slender with an oblong head, broad at the back with the eyes directed slightly upwards. Tail relatively thin, accounting for a fifth of the total length. Upperside grey-brown or olive-brown, with dark square markings which are usually arranged in 5 longitudinal rows. There is usually a V-shaped mark on the nape. The underside is whitish: yellowish or reddish, marked with irregular square black patches.

Distribution: Found in parts of central France, in the area around the Moselle, Rhine, Lahn and Nahe, and at Meissen in Saxony. Also in north and central Italy, Czechoslovakia, Switzerland, Austria, Hungary, Romania and many parts of the Balkans and south Russia. Occurs over large areas of Asia to north-west India and western China.

Distribution: The Iberian Peninsula, France, south-western Switzerland, north-west Italy, the Balearics, Corsica, Sardinia and Sicily. Also in north-west Africa.

Habitat and habits: Lives in the vicinity of large lakes and ponds and shelters under rocks, moss and piles of sticks. They can lie in the water and snap up fish that swim by, swallowing them alive. They also feed on frogs and earthworms and sometimes eat toads and small newts. Now and again they appear in large numbers in fishponds, and may do considerable damage. They mate in March–April. The 10–20 eggs are laid in June–July and hatch in August–October.

75 Dice snake
  Natrix tesselata

Identification: 100–120 cm (40–48 in.), exceptionally 150 cm (60 in.).

Habitat and habits: More closely associated with water than its relatives,

living mainly along the banks of streams, rivers, lakes and ponds. Scarcely comes on land except to bask and to lay eggs, but it does spend the winter on land. The 5–25 eggs are laid in loose soil or under leaves and stones in July–August.

Dice snakes dive and swim remarkably well. They feed mainly on fish, but also take frogs, tadpoles and aquatic newts. They either hunt their prey actively or lie on the bottom and wait for it to swim by.

76 Grass snake
Natrix natrix

Identification: 70–150 cm (27–60 in.), exceptionally longer; a specimen measuring 205 cm (82 in.) has been found on the Yugoslav island of Veglia. Naturally the largest individuals are found in places with a plentiful supply of food and the least danger from enemies. The long, slender body is slightly compressed, and the head is distinctly set off from the body, particularly in the juveniles. The flat snout is rounded, and the relatively large eyes have a black iris with a pale ring round the circular pupil. The rear end of the body continues without any transition zone into the pointed tail which accounts for a fifth of the total length. The scales on the body are rhomboid and largest on the sides, and they have a sharp and conspicuous longitudinal keel. The broad transverse ventral plates are specially developed scales which can be raised by muscles to form a series of sharp ridges that enable the snake to get a grip on the ground when moving. The coloration and pattern are very variable and depend upon age, sex and locality; in Europe upwards of 10

geographical forms have been recognized. The upperside is usually blackish-brown but may be pale or dark grey, olive or reddish. There are normally 2 longitudinal rows of large or small dark markings along each side of the body. On the upperside of the head the front part is always a uniform black, and the scales along the upperside of the mouth are whitish-yellow with black eges. The lower jaw is pale whitish, a colour which continues behind the corners of the mouth and behind the nape as a collar which is broken in the mid-dorsal line. When seen from above this broken band appears as the 2 half-moon-shaped markings which are so characteristic of the Grass snake. Occasionally these patches do meet on the mid-line, thus forming a true collar. The colour of the collar patches varies from white through yellow to orange, sometimes they are very conspicuous, and exceptionally they may be completely absent. To the rear these patches have a distinct broad black border. The throat and the front part of the underside are whitish but farther back there is an increasing amount of black.

The young resemble the adults, both in colour and pattern, but the neck patches are more conspicuous and have sharper borders. It has been said that one can tell the sex of a Grass snake by the colour of the neck patches, those of the female being more orange, but this is not the case. It is, however, correct that the largest specimens are females and that the tails of the females are slightly shorter and a little more pointed than those of the males.

Distribution: Throughout the whole of Europe, except Iceland, Ireland and

that part of Scandinavia which lies north of 67° N.

Habitat: Usually in the vicinity of standing water, in moorland and along the banks of rivers and lakes, often in places with a dense growth of rushes. It is thus well placed to find frogs which form the main item in its diet. It is therefore catastrophic for Grass snakes when such areas are drained, and no compensation that when dried out they will support mice, for the Grass snake is primarily a diurnal animal whereas mice are essentially nocturnal. Furthermore, not all Grass snakes will actually take mice. Large drainage schemes have therefore brought about a certain decrease in the populations of this snake; this is scarcely due to proximity to man because Grass snakes often live quite closely to houses and farms; and sometimes even in large gardens and parks. Like most other reptiles, the

Grass snake remains closely tied to its own home range, and only leaves it to hunt for prey.

Habits: An excellent swimmer, which bends the body from side to side, usually with the head above water. It can also lie and float at the surface, using its lungs as swim-bladders. When setting off to swim it fills the lungs with air which is exhaled when it submerges. It is very occasionally seen in salt water, sometimes quite far from land. It can remain submerged for more than an hour, and it is not at all uncommon for it to dive and remain under the water for 15–30 minutes. When it is out after food it does not lie and wait for the prey but hunts actively, the eyes being the most important sense organ. It seldom observes prey animals before they begin to move. Frogs are the principal prey and these are caught by the head or by one of the hind-legs. If it catches a frog by its flank it pushes its mouth sideways along the body until it has the head. At first, the frog tries to free itself, but it soon becomes still and usually slides down the throat without a sound, although occasionally a frog may make a frightful screech. Among the anurans the Grass snake prefers the Common frog and the European tree-frog, but it also takes Edible frogs and small newts, and there are even records of Grass snakes eating toads. It is not unusual for a large Grass snake to swallow 4 or 5 frogs in succession, but in such cases it will not need to feed again for several weeks. When frightened or disturbed it nearly always tries to escape. If cornered it will coil its body, raise its head and hiss, but only exceptionally will it bite. When seized in the hand a Grass snake

141

makes strenuous efforts to escape and produces a shiny yellow secretion from sac-shaped glands at the base of the tail. This secretion can be squirted out with considerable force, but usually it merely flows down the snake, giving off a distinctive smell. The normal function of this secretion is probably to oil the skin so that it becomes water-repellent. When disturbed and even if not seized a Grass snake will sometimes twist and turn and finally become completely smeared with the secretion. It also tries to avoid danger by holding the body rigid and lying motionless with open mouth, sometimes on its back, for up to half an hour. When seized by the end of the tail a Grass snake, unlike an Adder, can coil itself upwards and wind itself round one's arm.

During October, most Grass snakes leave their damp summer abodes and go into hibernation in more elevated places, such as stone walls, under tree stumps and in compost heaps. Often several of these snakes will hibernate together. The larger individuals enter hibernation 14 days or a month before the smaller ones and the young, and emerge last in the following spring. At the beginning of April the sunshine and the spring warmth entice Grass snakes out of hibernation, and after they have lain and basked for a couple of days they start to mate, although this may not occur until May or June. During mating the male and female lie side by side and twist their bodies and tails around each other. The female lays her eggs in the middle of the summer, in July–August, often in middens, piles of leaves or compost heaps, places where it is relatively humid and where rotting produces a certain

amount of warmth. The 10–20 grey ish-white oval eggs are 2–3 cm ($\frac{3}{4}$– in.) long and sticky; at first the shell i tough and leathery but it becomes har der and completely dry with time Large, old snakes may lay clutches o up to 40 eggs. Frequently several egg will stick together, although they ar laid separately with an interval of a few minutes or even half an hour be tween each one, so the adhesion of the eggs is something secondary. Since places suitable for egg-laying are often few and far between, one often finds the Grass snakes of an area congregat ing to lay in the same spot, so that there may be hundreds of eggs within a limited area. Embryonic develop ment starts within the female, so that the newly laid eggs may contain well developed young. There is consider able variation in the amount of embry onic development that takes place before the eggs are laid, and the subse quent development is much influenced by local climatic and temperature con ditions, so it is not really possible to give any accurate figures for the period of incubation. With plenty of sunny days and warmth from the rotting vegetable matter the eggs may hatch in 5 weeks, but there are records of peri ods twice as long as this. As a rule the young hatch at the beginning of September after they have torn their way out of the egg with the little egg tooth on the snout. The young are then black and 15–18 cm (6–7 in.) long. They can immediately fend for them selves, taking young fish, earthworms and slugs. On the other hand, they are at first unable to swim, possibly because they are not yet able to master the necessary breathing technique. The young which are born at the end of autumn go into hibernation immediately and if they survive the winter on an empty stomach they will find plenty of food in the form of tadpoles when they wake up in the following spring.

Grass snakes become sexually mature when 3–4 years old and about 50–60 cm (20–23 in.) long. They slough 5–6 times in the course of the summer. Before each sloughing they become sluggish and stop feeding, possibly because they are more or less blind because the epidermal covering of the eyes becomes opaque before the moult. The skin splits along the snout and it often comes off in one piece, the snake rubbing itself against roots, stems and stones. Sloughing is facilitated by a thin layer of cells which becomes slimy and this lubricates the inside of the old skin, so that it almost slides off. The slough itself is colourless and is of course inside out. It has stretched so much that one has to subtract about 15 per cent of its length to get an idea of the length of the snake from which it came.

Grass snakes have many enemies. The eggs are taken by rats, stoats and other small mammals which also feed on the newly hatched young. The adults' most dangerous enemy is man, for many people kill them in the mistaken belief that they are dangerous—or merely because they are snakes.

77 Hooded snake
Macroprotodon cucullatus

Identification: 40–50 cm (15–20 in.). Body slender with an elongated head, and somewhat similar in general appearance to the Smooth snake. The popular name is derived from the dark hood-like patch on the back of the head. The upperside is pale brown with small dark brown patches

arranged in rows down the back; these are larger on the sides than along the mid-line of the back. The underside is pale yellow or coral-red with black markings which sometimes fuse to form a dark central line. Occasionally the underside is uniformly pale.

Distribution: Found in the southern part of the Iberian Peninsula, in the Balearic Islands and in northern Africa.

Habitat and habits: Spends the day sheltering under a flat stone or in a hole in the ground, emerging at night to hunt for lizards, both those which are active at this time and also those which are lying up for the night.

78 Montpellier snake
Malpolon monspessulanus

Identification: 100–200 cm (40–80 in.), the largest individuals being found in the eastern part of the distribution range. Body long and slender, and head only slightly separated from the neck. On the upperside of the head there is a well-marked groove running from the eyes to the tip of the snout; this is most conspicuous in older specimens. The ground colour of the upperside is variable and may be pale sandy-brown, grey-brown, olive-green or almost completely black. Along the back and tail there are small black markings, often with white edges. These spots form 5 or even 7 more or less distinct longitudinal rows, which are most conspicuous in young individuals.

Distribution: In the Iberian Peninsula, southern France and Liguria, eastwards through the Balkans to Macedonia and southern Bulgaria.

Also in north-west Africa, Asia Minor, and farther east to Iran and the Caucasus.

Habitat and habits: Lives in sunny and dry, stony places with low bushy vegetation and usually keeps well hidden. If, however, it is approached it will betray its hiding-place by hissing loudly. The diet consists of rats and other small mammals, birds and especially lizards. It also eats snakes and grasshoppers. Normally it does not lie in wait for its prey, but hunts actively, often with its mouth half-open. It is venomous, and the bite normally kills the prey in about 3 or 4 minutes. If the prey is large and powerful, the snake will usually coil round it to hold it down, but it does not try to constrict it. It is evidently not dangerous to man, and this is because the venom fangs are placed so far back in the mouth, that they cannot be brought into play when the snake tries to bite

a human being. The female lays 4–12 eggs at the end of July, and the young hatch in October.

79 Snub-nosed viper
Vipera latastei

Identification: 50–60 cm (20–23 in.). Head clearly separated from the neck and body. Tip of snout short, soft and upturned. Upperside greyish, becoming brownish in the male and reddish-brown in the female. Running along the back there is a wavy band of fused rhomboid or irregular dark markings. Underside yellowish or grey with dark spots or marbled markings.

Distribution: In Europe only in Portugal and Spain, but also occurs in Morocco, Algeria and Tunisia in north Africa.

Habitat and habits: Lives in river valleys, extending up into the mountains. In general, its habits are similar to those of the Adder.
   Also known as Lataste's viper.

80 Adder
Vipera berus

Identification: 50–60 cm (20–23 in.), females sometimes longer. Body very stout and thick-set. Head flat, broad at the back and clearly separated from the body, and shortest and broadest in the female. The tail is short, and its tip may be yellow, orange or red; it accounts for only a sixth of the total length in the male, an eighth in the female. In the male the front region of the tail is thickened; this is due to the presence of the double copulatory organ. The eye has a reddish iris and above it the scales protrude to form a kind of eyebrow. The narrow black tongue is forked at the tip, and it can be shot in and out without the mouth being opened, for as in other snakes there is a small gap between the lips through which the tongue can slip. The horny scales on the body nearly all have a distinct longitudinal keel and they are arranged in longitudinal rows. The scales are oblong and increase in size from the back towards the belly. The broad, overlapping ventral plates extend from side to side. The colours and patterns vary according to sex, age, locality, season and the stage of moulting or sloughing. The Adder is one of the few snakes that shows sexual dimorphism in colour, the pale greyish specimens with black markings being males, the brownish ones with less prominent markings being the females or juveniles; adders are usually born brown and only assume their definitive coloration after 2 or 3 years. There are, of course, all kinds of intermediate colours, and indeed the upperside may be anything from pale grey or pale yellow to red-brown or black, occasionally even white. The colours are always most vivid after the first slough of the year. At the rear of the head there is a dark marking in the form of a cross or a V. Along the edge of the upper jaw the horny plates are usually white edged with black. Starting from the back of the head there is usually a zigzag stripe which continues down along the mid-line of the back and right out to the tip of the tail. This stripe consists of a row of irregular, more or less continuous markings. On each side of the body there is nearly always a longitudinal stripe, which is normally broken up into spots of varying sizes. The dark markings are particularly conspicuous in individuals with a pale ground colour,

but sometimes they are difficult to distinguish and some specimens are so dark that one can only catch a faint glimpse of the pattern. There is also a jet-black variety in which the dorsal stripe and the other markings have completely disappeared. The young of this black variety are born with colours and markings and they do not become completely black until they are about 2 years old. Occasionally specimens occur which are a uniform red-brown.

Distribution: The Adder is more widely distributed than any other European snake. It is found in all the temperate parts of Europe and right across Asia to the Pacific Ocean. The southern limits run from northern Spain and Portugal through central Italy, Yugoslavia, Romania and the Ukraine to the southern Urals, Altai and Vladivostok. In the north it occurs in Scotland, Lapland, Finland and

northern Russia. It is, however, un-
known in Iceland, Ireland and the
islands of the Mediterranean.

Habitat: Lives mainly on moorland,
heaths and sand dunes, in stone walls
and on the outskirts of woodland
where it can bask in the sun. In mountainous regions it extends up to altitudes of 3,000 m (9,800 ft).

Habits: Often retreats into mouse burrows, and seldom lives in the vicinity
of water, although it swims very well
and can sometimes be seen far out in a
lake among the reeds. Adders are and
have been much persecuted by man,
who not only kills them but also affects
them indirectly by his agricultural
activities.

Adders start to emerge from hibernation at the end of March or the
beginning of April, depending on
whether the day temperature rises
about 10°C and the sun starts to warm
the ground. During the summer they
are easiest to find between five and
eight o'clock in the morning and
eight and ten o'clock at night. But
even though they are essentially diurnal animals they still retain the vertical pupil characteristic of many
nocturnal animals. The explanation
for this seems to be that in the
southern and warmer parts of their
range they are, in fact, crepuscular
or nocturnal. In the north they must
first be warmed up by the morning
sun before they can go hunting and
subsequently digest their food. When
the days become shorter and the
nights cold in September–October
they go into hibernation in holes in the
ground or under tree stumps. As
moorland and heaths offer rather poor
hiding-places, these snakes sometimes
have to crawl for long distances before

finding suitable places. In fact, one
often finds several in the same place,
and aggregations of 20–30 are not
uncommon.

Adders feed mainly on mice,
lizards, Slowworms and frogs,
although they also take the eggs and
young of small birds. They may eat
20–30 mice during the course of a
summer; sometimes they creep into a
mouse burrow and eat a whole litter of
mice and the mother. Digestion is slow
and localized. The first item of prey
swallowed may be almost digested
before the enzymes start to work on
the remainder. As soon as they
become aware of prey in the vicinity
they orientate themselves by darting
the tongue in and out, thus carrying
scent particles to the organ of
Jacobson in the palate. When the prey
has come close enough they strike so
rapidly that it is almost impossible to
see what happens. The mouth is
opened wide and the venom fangs are
directed forwards. Each side of the
upper jaw has one hollow fang, about
4–5 mm ($\frac{1}{5}$ in.) long, which is only in
use for a few months, after which it
falls out and is replaced by a reserve
fang from the back of the mouth.
When the mouth is shut the tips of the
venom fangs are turned backwards
and protected by a fold of skin. When
about to strike it therefore has to lift
the whole of the upper jaw upwards so
that the fangs assume the right position for striking. Adder venom is a
clear yellowish secretion which acts
mainly on the blood system. A single
bite may release about 0·1 g ($\frac{1}{300}$ oz) of
venom.

At the end of April or the beginning
of May the males fight violently over
the females. When two males meet
they will often carry out what is known

as the 'dance' of the Adders. They raise the front parts of their bodies and start pushing against each other, but no blood is shed and the fangs are not used. Mating usually takes place in a sunny place, and sometimes several Adders will lie intertwined in the same place. The young are born in August or September. The number born depends on the female's age but is usually 5–15, although occasionally it may be 20 or more. The young Adders, about 14–21 cm ($5\frac{1}{2}$–$8\frac{1}{2}$ in.) in length, are enclosed in a thin elastic egg membrane which is burst immediately after birth. They normally slough very shortly after birth. The fangs are functional at this time so that even newly born young can feed themselves, although they do not, in fact, have much time for this before starting their first period of hibernation.

Adders grow slowly and are not sexually mature until they are 4–5 years old and about 50 cm (20 in.) long. It is not known how long they can live. They have many enemies, including large birds of prey, crows and large gulls, and also hedgehogs. A hedgehog will uncoil and give the snake a bite and then coil itself up, presenting a thicket of spines. This will be repeated several times until finally the Adder is killed. The hedgehog then moves along the body of its prey, crushing the bones as it goes, before eating it. The Adder is dangerous to man but only exceptionally causes death; it has been accused of lying in wait to attack humans but this is wrong and it only strikes to obtain food or to defend itself. Adder bites can be treated by the injection of a special anti-venine. Before this was introduced, treatment was primitive,

involving tourniquets, suction of the wound, cauterization and lancing, or taking alcohol. As the venom spreads through the bloodstream a tourniquet above the bite should stop the venom spreading, but this does not always happen and a tourniquet left on for too long may result in gangrene of the tissues. Suction of the wound is not necessarily of any use to the patient. It has often been said that the wound should be burnt or lanced so that the venom can flow out with the blood, but this method is evidently not satisfactory, and it also increases the chances of infection. Nor is alcohol to be recommended as it increases blood flow and thus the absorption of the venom, which in the Adder is of course primarily a blood poison. Treatment with anti-venine is the only method which has proved effective. But more recently it has been recommended that the patient should be kept quiet and transported as rapidly as possible to a hospital for observation. If there is no local reaction within $\frac{1}{2}$–1 hour after the bite, treatment is unnecessary. If there is a local reaction but no symptoms of poisoning within 2 hours the patient should be kept under observation for up to 6 hours after the bite, but anti-venine treatment appears to have no effect on the course of a local reaction. With severe symptoms of poisoning involving difficulties in breathing the patient should be given a blood transfusion and oxygen and ventilated by respirator; if necessary, tracheotomy (opening of the trachea) should be performed and an intravenous injection of Adder anti-venine should be given.

Also known as the Northern or European viper.

81 Meadow viper
Vipera ursinii

Identification: 40–55 cm (15–22 in.). A small viper with an almost triangular head and a pointed snout. The tail is short, accounting for only an eighth to a twelfth of the total length. The ground colour of the upperside is yellowish, brownish or olive-green, the females being mainly brownish. Along the mid-line of the back and tail there is an irregular, more or less continuous wavy band which may be dark grey, red-brown or blackish-brown and is reminiscent of the Adder's zig-zag stripe. In addition, each side of the body is usually marked with two rows of black or brown spots, those closest to the belly being the smaller. Sometimes these spots fuse to form longitudinal bands. The patterns become less distinct with age. The underside is yellowish, whitish or greyish with dark marbling.

Distribution: There are several large or small isolated populations in central and eastern Europe, ranging from south-east France, central Italy through Austria, Hungary, Yugoslavia, Romania, Bulgaria to south Russia and the Caucasus, and farther east into central Asia.

Habitat and habits: Lives mainly in low-lying country, in meadows and heathland, but extending up into the mountains in some parts of the range. Often shelters in the burrows of mice and other small mammals. Feeds almost exclusively on insects, lizards and sometimes small mammals. The bite is only slightly venomous and it usually swallows its prey living.

Also known as Orsini's viper and Field adder.

82 Asp viper
Vipera aspis

Identification: 50–60 cm (20–23 in.), the male exceptionally up to 75 cm (29 in.). The head is distinct from the neck and the tip of the snout is upturned. The ground colour of the upperside varies considerably and may be yellowish, reddish-yellow, reddish-brown or grey. Along the upperside of the body and tail there is a series of narrow black transverse bands, which in some cases may fuse to form a zigzag band with rounded angles. The underside may be uniform yellowish, greyish or almost completely black, with or without markings. The underside of the tip of the tail is sulphur-yellow or orange-red.

Distribution: Pyrenees, southern France, southern Black Forest, the

Alps, Italy, where it is the commonest venomous snake, and a few places in Yugoslavia. Not found outside Europe.

Habitat and habits: Prefers dry hilly country and mountains, particularly limestone rocks, but seldom reaches altitudes above 2,000 m (6,540 ft).

The Asp is active both by day and night, is not very aggressive and moves about relatively slowly. The diet consists of mice, lizards and sometimes nestling birds. The juveniles take insects and worms. Asps emerge from hibernation at the end of March or the beginning of April and mate 2–3 weeks later. After a gestation period of 4 months the female gives birth to 5–15 young are 18–20 cm (7–8 in.) long. The venom from an Asp bite would produce considerable pain in man, but is not normally fatal.

83 Sand viper
Vipera ammodytes

Identification: 60–80 cm (23–30 in.), the males occasionally up to 90 cm (35 in.). The snout has a soft, scaly tip, but otherwise this species is rather similar to the Adder and the Meadow viper in general coloration and in possessing a zigzag dorsal stripe. The broad, triangular head is distinctly separated from the neck. The body is very stout and the short tail is thin and pointed. The ground colour of the upperside is yellow-brown, sandy-grey, grey-brown or brown. Along the mid-line of the back there is a dark, often wavy or zigzag band, which sometimes has a black border, and which is darkest and most conspicuous in the males. The underside is yellow-brown with grey-brown or blackish spots. The underside of the tip of the tail is sulphur-yellow or brick-red.

Distribution: North-eastern Italy, southern Tyrol, Istria, Austria, Hungary, Yugoslavia and other parts of the Balkans. Also in Asia Minor, Syria and Iran.

Habitat and habits: Lives in dry, rocky places with scattered vegetation. Mainly active at night, normally spending the day under boulders, in ruins or in thick bushy undergrowth. Sometimes, however, the Sand viper climbs up into bushes and basks in the sun. In mountainous areas it reaches altitudes of up to 2,000 m (6,540 ft).

This is the largest and most dangerous venomous snake in Europe, but fortunately it normally makes its presence known by hissing when one approaches it. The hiss starts faintly but becomes louder the closer one approaches.

They feed mainly on mice and voles, sometimes also on small birds, snakes, Slowworms and lizards. The juveniles at first feed on young lizards and evidently also take invertebrates.

Also known as the Long-nosed viper.

BIBLIOGRAPHY

Appleby, L. G., *British Snakes*, Morrison & Gibb Ltd., London and Edinburgh, 1971

Hvass, M., *Reptiles and Amphibians of the World*, Methuen and Co., London, 1964

Mertens, R., and Wermuth, H., *Die Amphibien und Reptilien Europas*, Verlag Waldemar Kramer, Frankfurt am Main, 1960

Simms, C., *Lives of British Lizards*, Goose and Son, Norwich, 1970

Smith, M., *The British Amphibians and Reptiles*, 3rd edition, Collins, London, 1964

Steward, J. W., *The Tailed Amphibians of Europe,* David and Charles, Newton Abbot, 1969

Steward, J. W., *The Snakes of Europe,* David & Charles, Newton Abbot, 1971

Vogel, Z., *Reptiles and Amphibians*, Studio Vista, London, 1964

INDEX

English names

The figures refer to both the illustrations and the descriptions.

Scientific names

The figures refer to both the illustrations and the descriptions, except that where a page number is given this refers to one of a few reptiles that are not illustrated.